Moritz A. Frenzel

Advanced Structural Finite Element Modeling

Moritz A. Frenzel

Advanced Structural Finite Element Modeling

of Arterial Walls for Patient-Specific Geometries

Südwestdeutscher Verlag für Hochschulschriften

Impressum/Imprint (nur für Deutschland/ only for Germany)

Bibliografische Information der Deutschen Nationalbibliothek: Die Deutsche Nationalbibliothek verzeichnet diese Publikation in der Deutschen Nationalbibliografie; detaillierte bibliografische Daten sind im Internet über http://dnb.d-nb.de abrufbar.

Alle in diesem Buch genannten Marken und Produktnamen unterliegen warenzeichen-, marken- oder patentrechtlichem Schutz bzw. sind Warenzeichen oder eingetragene Warenzeichen der jeweiligen Inhaber. Die Wiedergabe von Marken, Produktnamen, Gebrauchsnamen, Handelsnamen, Warenbezeichnungen u.s.w. in diesem Werk berechtigt auch ohne besondere Kennzeichnung nicht zu der Annahme, dass solche Namen im Sinne der Warenzeichen- und Markenschutzgesetzgebung als frei zu betrachten wären und daher von jedermann benutzt werden dürften.

Verlag: Südwestdeutscher Verlag für Hochschulschriften Aktiengesellschaft & Co. KG
Dudweiler Landstr. 99, 66123 Saarbrücken, Deutschland
Telefon +49 681 37 20 271-1, Telefax +49 681 37 20 271-0
Email: info@svh-verlag.de
Zugl.: München, TU, Diss., 2009

Herstellung in Deutschland:
Schaltungsdienst Lange o.H.G., Berlin
Books on Demand GmbH, Norderstedt
Reha GmbH, Saarbrücken
Amazon Distribution GmbH, Leipzig
ISBN: 978-3-8381-1404-0

Imprint (only for USA, GB)

Bibliographic information published by the Deutsche Nationalbibliothek: The Deutsche Nationalbibliothek lists this publication in the Deutsche Nationalbibliografie; detailed bibliographic data are available in the Internet at http://dnb.d-nb.de.

Any brand names and product names mentioned in this book are subject to trademark, brand or patent protection and are trademarks or registered trademarks of their respective holders. The use of brand names, product names, common names, trade names, product descriptions etc. even without a particular marking in this works is in no way to be construed to mean that such names may be regarded as unrestricted in respect of trademark and brand protection legislation and could thus be used by anyone.

Publisher: Südwestdeutscher Verlag für Hochschulschriften Aktiengesellschaft & Co. KG
Dudweiler Landstr. 99, 66123 Saarbrücken, Germany
Phone +49 681 37 20 271-1, Fax +49 681 37 20 271-0
Email: info@svh-verlag.de

Printed in the U.S.A.
Printed in the U.K. by (see last page)
ISBN: 978-3-8381-1404-0

Copyright © 2010 by the author and Südwestdeutscher Verlag für Hochschulschriften Aktiengesellschaft & Co. KG and licensors
All rights reserved. Saarbrücken 2010

Abstract

Cardiovascular diseases including atherosclerosis and aneurysm are the leading cause of human mortality in the western world. The interrelation of the onset and development of such diseases with mechanical forces acting on and within the arterial wall represents a research field of eminent interest. The arterial wall is a highly complex tissue characterized by a sophisticated microstructure. Its structural response exhibits strong nonlinearity, reflected by large deformations and strains, material inhomogeneity, anisotropy and viscoelasticity. Moreover, it is a living structure which permanently adapts to its physical environment, phenomena known as growth and remodeling.

This dissertation is concerned with a finite element approach that is suitable to model the arterial wall with all its complexities. A recent advance in biomechanical research is the mechanical simulation of patient-specific geometric models acquired by medical image technology with the aim of studying inter-individual differences to further understand vascular diseases. The simulation of such patient-specific arterial wall structures is computationally highly demanding. Challenges stem from the inevitably three-dimensional modeling and the advanced material characteristics of the wall.

Standard three-dimensional elements are known to reflect poor performance within incompressible and bending-dominated problems. A set of efficient and accurate elements is proposed, including hexahedrons, 8-node solid-shells and wedge-shaped solid-shells, which take the thin-walled shape and incompressibility of the arterial wall into account. To tackle various locking phenomena different methods are reviewed and combined and their performance is demonstrated. In addition, constitutive laws which consider characteristics like incompressibility, anisotropy and viscoelasticity for the arterial wall are examined. This includes popular anisotropic laws based on structural tensors, a microstructurally motivated anisotropic Continuum Chain Network model, and the extension to viscoelasticity. Next, recent remodeling approaches in terms of fiber reorientation are presented. Two different approaches are implemented and evaluated in examples such as idealized arterial vessels and bifurcations.

Finally, these three methodologies, advanced element technology, sophisticated constitutive laws, and remodeling approaches are combined within our in-house software framework to analyze patient-specific wall structures. The application to these geometries requires the generation of the wall which is not available from medical imaging. Enhancing modeling strategies are employed based on the vessels' centerline to generate a physiologically varying wall thickness determined by the local diameter and a realistic winding of the fiber pattern along the vasculature. Combined with subsequent remodeling, a considerably realistic arterial wall model is obtained as demonstrated by two examples containing the aortic arch and the iliac bifurcation, each with branching vessels.

The proposed approach is envisioned to enable multiple further patient-specific simulations. The resulting data could then be correlated with clinical and experimental findings with the aim of further deepening the understanding of vascular diseases.

Acknowledgements

A number of people have contributed to this work and supported me in different ways. I would like to thank them here and most cordially:

Prof. Dr.-Ing. Wolfgang Wall for supervising me, Prof. Dr.-Ing. Jörg Schröder for being second referee, Prof. Dr. mont. Ewald Werner for chairing the examination board, Prof. Dr.-Ing. Manfred Bischoff for teaching me finite element technology, Dr. Burkhard Bornemann for teaching me nonlinear continuum mechanics, Dr. Andrew Comerford, Thomas Klöppel and Markus Gitterle for proofreading the manuscript, and all other colleagues at the Institute for Computational Mechanics for enjoyable lunch hours.

Contents

Glossary   vii

1 Introduction to Arterial Wall Mechanics   1
  1.1 Constitution of the Artery and Related Diseases 1
      1.1.1 Arterial Wall Layers 2
      1.1.2 Common Arterial Wall Diseases 4
  1.2 Structural Characteristics of the Arterial Wall 5
  1.3 Review of Computational Modeling in Vascular Mechanics 7
      1.3.1 Patient-Specific Modeling 7
      1.3.2 Computational Hemodynamics 8
      1.3.3 Fluid-Structure-Interaction Models 9
      1.3.4 Advanced Structural Wall Models 11
  1.4 Objectives of the Present Work 13

2 Some Nonlinear Continuum Mechanics   17
  2.1 Kinematics 17
  2.2 Stress Concept 20
  2.3 Balance Principles and Entropy 21
  2.4 Initial Boundary Value Problem 23

3 Efficient Finite Elements for Arterial Wall Modeling   25
  3.1 Prerequisites of Finite Element Analysis 25
      3.1.1 Variational Principles 26
      3.1.2 Linearization of the Virtual Work Principle 28
      3.1.3 Discretization of the Virtual Work Principle 30
  3.2 Locking Phenomena of 3D Elements 34
      3.2.1 Volumetric Locking 35
      3.2.2 Shear Locking 37
      3.2.3 Trapezoidal Locking 38
      3.2.4 Membrane Locking 39
  3.3 Limits of Element Perfectibility and the Issue of Mesh Quality 39
      3.3.1 Element Perfectibility — MacNeal's Dilemma 40
      3.3.2 Influence of Mesh Quality 41

	3.3.3 Locking in Large Deformation Analyses 42
3.4	Methods to Eliminate Locking Phenomena 43
	3.4.1 Assumed (Natural) Strain Method 43
	3.4.2 The Discrete Strain Gap Method 45
	3.4.3 The Enhanced Assumed Strain Method 49
	3.4.4 Alternative Methods 52
3.5	Efficient 3D Finite Element Formulations 54
	3.5.1 A Bulky Hexahedral Element 54
	3.5.2 A Hexahedral Solid-Shell Element 57
	3.5.3 A Wedge-Shaped Solid-Shell Element 62
3.6	Numerical Benchmark Examples 66
	3.6.1 Slit Annular Plate 67
	3.6.2 Pullout of Open-ended Circular Cylinder 69
	3.6.3 Pinched Hemisphere 71
	3.6.4 Cook's Membrane Problem 74
	3.6.5 Summary 75

4 Constitutive Models for the Arterial Wall   77

4.1	Essentials of Constitutive Modeling 77
	4.1.1 Properties and Restrictions for Constitutive Laws 77
	4.1.2 Constitutive Equations in Terms of Invariants 79
	4.1.3 Incompressibility and Near Incompressibility 80
	4.1.4 Extension to Anisotropy 83
4.2	Basic Isotropic Models 87
	4.2.1 Neo-Hookean Material 87
	4.2.2 Mooney-Rivlin Material Law 87
4.3	A Family of Anisotropic Models 88
	4.3.1 Holzapfel's Model 88
	4.3.2 Balzani's Model 90
	4.3.3 Modified Anisotropic Models 91
4.4	Continuum Molecule Chain Models 93
	4.4.1 Mechanical Response of a Single Fiber 94
	4.4.2 Chain Network Model 97
4.5	Extension to Viscoelasticity 100
	4.5.1 Linear Generalized Maxwell-Model 101
	4.5.2 Large Strain Fiber-Reinforced Viscoelasticity 103
4.6	Numerical Examples 106
	4.6.1 Stretching of a Rubber Sheet 106
	4.6.2 Plate with Differing Warp and Fill 108
	4.6.3 Inflation of a Fiber-Reinforced Rubber Tube 108
	4.6.4 Cyclic Inflation of a Viscoelastic Fiber-Reinforced Tube 110

5 The Biomechanical Phenomenon of Remodeling 113
  5.1 Literature Review and Definition of Terms 113
  5.2 Governing Equations 115
  5.3 Fiber Remodeling and Review of Recent Approaches 117
    5.3.1 Driessen's Approach 117
    5.3.2 Hariton's Approach 119
    5.3.3 Kuhl's Approach 121
  5.4 Suggested Advancements of Hariton's and
      Kuhl's Remodeling Approaches 123
  5.5 Numerical Examples 126
    5.5.1 Idealized Human Carotid Artery Model 126
    5.5.2 Idealized Human Carotid Bifurcation Model 129
    5.5.3 Idealized Tendon Model 131
    5.5.4 Idealized Artery Model 133
  5.6 Discussion of the Presented Fiber Remodeling Approaches 136

6 Simulation of Patient-Specific Arterial Wall Models 141
  6.1 Wall Model Generation 141
    6.1.1 Segmenting and Processing CT-Data 142
    6.1.2 Generating the Wall Geometry 142
    6.1.3 Integrating Centerline Data to Enhance the Wall Model 145
    6.1.4 Considering the Initial Wall Stress State 148
  6.2 Simulation of a Patient-Specific Aortic Arch 150
    6.2.1 Problem Setup and Reference Parameters 151
    6.2.2 Influence Study 152
  6.3 Simulation of a Patient-Specific Section at the Iliac Bifurcations 155
  6.4 Discussion 156

7 Conclusion and Outlook 161

Glossary

Abbreviations

AAA	abdominal aortic aneurysm
ANS	Assumed Natural Strain method
BE	balance equation
CE	constitutive equation
CFD	computational fluid dynamics
CT	computer tomography
DBC	Dirichlet boundary condition
DSG	Discrete Strain Gap method
EAS	Enhanced Assumed Strain method
FBC	force (Neumann) boundary condition
FSI	fluid-structure-interaction
IA	intracranial aneurysm
KE	kinematic equation
MRI	magnetic resonsnce imaging
PvW	principle of virtual work
RHS	right-hand-side
SMC	smooth muscle cell
VHW	Fraeijs de Veubeke-Hu-Washizu principle
WLC	wormlike chain model

Operators

$(\dot{\,\cdot\,}) = \frac{\mathrm{d}}{\mathrm{d}t}(\,\cdot\,)$	material time derivative
cof	cofactor
D	directional derivative (or Gâteaux derivative)
Div	divergence w.r.t. material system $\mathbf{X}$
dev	deviator operator
div	divergence w.r.t. spatial system $\mathbf{x}$
$\mathrm{Grad}(\,\cdot\,)$	gradient w.r.t. material system $\mathbf{X}$
$\mathrm{grad}(\,\cdot\,)$	gradient w.r.t. spatial system $\mathbf{x}$
Lin	linearization

tr	trace operator
Δ	linearization increment
δ	virtual operator
ϵ	infintesimal eps
∇_0	Nabla operator

Nomenclature

t	time
$\mathcal{B}_0$	reference configuration
$\mathcal{B}_t$	current configuration
$\mathbb{R}^3$	three-dimensional space
$\mathbf{X}$	material points/coordinates
$\mathbf{x}$	spatial points/coordinates
φ_t	motion at fixed time t
$\mathbf{u}$	displacement field
$\mathbf{V}$	material velocity field
$\mathbf{v}$	spatial velocity field
$\mathbf{A}$	material acceleration field
$\mathbf{a}$	spatial acceleration field
$\mathbf{F}$	material deformation gradient
J	Jacobi determinant
$\mathbf{U}$	right (material) stretch tensor
$\mathbf{R}$	orthogonal rotation tensor
$\mathbf{v}$	left (spatial) stretch tensor
$\mathbf{C}$	right Cauchy-Green tensor
$\mathbf{b}$	left Cauchy-Green (Finger) tensor
$\mathbf{E}$	Green-Lagrangean strain tensor
$\mathbf{e}$	Euler-Almansi strain tensor
$\mathbf{L}$	material velocity gradient
$\mathbf{l}$	spatial velocity gradient
$\mathbf{t}$	surface traction force vector
$\mathbf{N}$	material surface normal
$\mathbf{n}$	spatial surface normal
$\boldsymbol{\sigma}$	Cauchy-stress tensor
$\mathbf{P}$	first Piola-Kirchhoff stress tensor
$\mathbf{S}$	second Piola-Kirchhoff stress tensor
$\boldsymbol{\tau}$	Kirchhoff stress tensor
p	hydrostatic pressure
ρ	spatial density
ρ_0	material density

Glossary

$\mathbf{b}_0$	volume force
e	internal energy per unit reference mass
r	external heat source per unit reference mass, Chapter 2
$\mathbf{q}$	spatial heat flux
$\mathcal{P}_{\text{int}}$	internal mechanical power
η	entropy
ϑ	absolute temperature
$\mathcal{D}$	dissipation per unit reference mass
ψ	(Helmholtz) free energy w.r.t. unit reference mass
Ψ	(Helmholtz) free energy w.r.t. unit reference volume
Π	potential energy
$\mathbb{C}$	elasticity tensor
$\mathcal{B}_{(e)}$	element domain
$\mathbf{d}$	nodal displacements
$\mathbf{N}$	matrix of shape functions, Chapter 3
$(\,\cdot\,)_{(e)}$	w.r.t. element domain
$(\,\cdot\,)^h$	finite element approximation
n_{ele}	number of element within domain
n_{nd}	number of nodes within domain
ξ, η, ζ	general/hexahedral element parameter space
r, s, t	triangular element parameter space
$\mathbf{J}$	Jacobi matrix mapping parameter space to material space
$\mathbf{B}$	B-operator matrix
$\mathbf{k}$	elemental stiffness matrix
$\mathbf{f}$	discrete force vector
$\mathbf{K}$	global tangent stiffness matrix
$\mathbf{D}$	vector of global unknown nodal displacements
ε	linear (small) strain tensor
E	Young's modulus
ν	Poisson's ratio
$\mathbf{I}$	identity tensor
$\mathbf{M}$	matrix of EAS ansatz, Chapter 3
$\boldsymbol{\alpha}$	EAS parameters, Chapter 3
$\mathbf{T}$	transformation matrix
$\mathbf{Q}$	orthogonal tensor
λ	principal stretch
I_1, I_2, I_3	principal invariants of $\mathbf{C}$
$\widehat{(\,\cdot\,)}$	isochoric contribution of $(\,\cdot\,)$ from multiplicative split of $\mathbf{F}$
$\mathbb{I}$	fourth-order identity tensor
$\mathbb{P}$	projection tensor

μ	shear modulus
κ	bulk modulus
Λ	Lamé parameter
$c_i, k_i, \mu_i, \epsilon_i, \alpha_i$	material parameters
$\mathbf{m}, \mathbf{n}$	unit vector field describing anisotropic fiber directions
$\boldsymbol{M}, \boldsymbol{N}$	structural tensors
$\bar{I}_i$	non-standard invariant i
γ	fiber angle w.r.t. circumferential direction
k	Boltzmann's constant, Section 4.4
N	number of chains w.r.t. unit reference volume, Section 4.4
r	molecule chain end-to-end length, Section 4.4
l	molecule bond length, Section 4.4
n	number of molecule bonds, Section 4.4
L	molecule chain contour length, Section 4.4
A	molecule chain persistence length, Section 4.4
$\mathcal{L}$	Langevin function, Section 4.4
β	Langevin function parameter, Section 4.4
$\mathbf{a}_i$	unit-cell material axes, Section 4.4
a_i	unit-cell material dimensions, Section 4.4
$\boldsymbol{Q}$	viscoelastic internal non-equilibrium stress, Section 4.5
$\boldsymbol{\Gamma}$	viscoelastic strain-like internal variable, Section 4.5
Υ	viscous configurational free energies, Section 4.5
τ	relaxation time, Section 4.5
β	free-energy factor, Section 4.5
$\boldsymbol{K}^r$	remodeling tangent map
$\boldsymbol{K}^c$	compatibility restoring tangent map
$\boldsymbol{\Sigma}$	configurational stress
$\boldsymbol{\mathcal{E}}$	Eshelby stress
ϱ	probability density function
Λ^2	squared mean of squared fiber stretch
ϕ^σ	Cauchy stress principal direction
λ^σ	principal Cauchy stress

1. Introduction to Arterial Wall Mechanics

This thesis is concerned with the arterial wall and a computational mechanics approach to assess its structural stress state with the help of nonlinear finite element methods. This includes advanced finite element technology to prevent well-known locking phenomena, state-of-the-art material modeling taking anisotropy, microstructure, and viscoelasticity into account, consideration of remodeling in the sense of mechanically motivated fiber reorientation, and a number of modeling strategies to generate a sophisticated, realistic and patient-specific arterial wall geometry. To set the stage, we will first give an overview of constitution, structural and disease characteristics of the wall, and subsequently review the literature on computational approaches to the topic. Finally, the objectives of the present work are summarized.

1.1 Constitution of the Artery and Related Diseases

The heart is the central pump of the cardiovascular system and a vast number of arteries carry the oxygen rich blood from the heart to the peripheral regions and the veins transport deoxygenated blood from these peripheral regions back to the heart. The arterial vessels are organized into an arterial tree, where the aorta as a single systemic artery emerges from the heart and successively branches into hundreds of arteries of progressively smaller caliber. The arteries thereby serve two functions, first they are conduits for blood flow and second they form a pressure reservoir. About three-quarters of the blood volume resides in veins at low transmural pressure and one-quarter resides in arteries at high transmural pressure.

To fulfill these functions at the high pulsatile pressure regime the arteries have to sustain considerable mechanical influence. Their structure is quite complex consisting of several layers with different functions and constituents. The main components of the vessel wall are: endothelium, smooth muscle cells, elastic tissue, collagen, and connective tissue. Associated with arterial branches and bifurcations macro- and microstructural changes can be observed, for instance a significant increase in thickness at branch ostium or bifurcation apex of arterial walls has been observed (Thubrikar [301]). Furthermore, the fibrous microstructure changes leading to additional strength of the wall at these locations. These regions seem to play a key role in vascular mechanics, not only from a hemodynamic point of view inducing flow perturbations, but also from a structural point of view bearing significant stress concentrations. Important sources of information about the biomechanics of the cardiovascular system are the books of Humphrey [152], Fung [104, 105], Thubrikar [301], and McDonald [210].

Collagen is the most abundant protein in mammals and also the ubiquitous load-bearing element in soft tissue (Fratzl [95, 96]). It confers mechanical stability, strength and toughness to a range of tissues from tendons and ligaments, to skin, cornea, bone and dentin. Its structural characteristic varies from highly elastic to brittle stiff. Collagen gains its versatility mostly by modifying its hierarchical structure, where the basic building block is the collagen fibril, a fiber with a thickness of $\sim 50-300$ nm. The other load-carrying protein in the arterial wall is elastin which is much more flexible than collagen and can sustain large stresses and strains. It shows an almost linear stress response over a large strain range.

Both collagen and elastin are synthesized by smooth muscle cells. They are the living components of the wall and control blood flow by vasoconstriction and vasodilatation. Furthermore, they are responsible for restoring the structural components during the remodeling process. Here, the term remodeling is understood as reorganization of existing and/or synthesizing new constituents such as collagen resulting in a different composition of the tissue. The term remodeling is used inconsistently in the literature and we refer to Chapter 5 for a detailed discussion of its definition. In the present work we focus on the passive response of the arterial wall. Active contributions by contraction of smooth muscle cells are thus not considered.

The arteries are arranged in three concentric layers, or tunics: the innermost tunica *intima*, the middle tunica *media*, and the outermost tunica *adventitia*, as described in the following, see also Fig. 1.1.

1.1.1 Arterial Wall Layers

Intima The intima consists of a monolayer of 0.2 to 0.5 μm thick endothelial cells held on a 1 μm thick basal membrane mostly made of collagen. The intima of a healthy artery has a negligible structural role because it is very thin. It is a nonclotting interface with blood and serves as gateway for molecule transport to and from the bloodstream. A variably thick internal elastic lamina usually serves as a prominent boundary between the intima and the media.

Media The media begins at the internal elastic lamina that lines the intima and extends to an external elastic lamina next to the adventitia. Both these laminae are perforated sheets of melded elastic fibers that permit the transmural transfer of water, nutrients, and electrolytes as well as cellular communication between adjacent arterial tunics. The external elastic lamina is less prominent in muscular arteries and does not exist in cerebral arteries.

The media consists of concentric patterns of elastic fibers and smooth muscle cells. In an elastic artery, the pattern consists of elastic lamellar units or muscular-elastic fascicles whose number and thickness vary along the vascular tree (up to 60, about 15 μm thick in the thoracic aorta; up to 30, about 20 μm thick in the abdominal aorta). Each unit can be seen as layers of smooth muscle cells separated by 3 μm-thick fenestrated sheets of elastic fibers forming a continuous fibrous helix. The pitch of the helix within the media is usually small leading to

1. Introduction to Arterial Wall Mechanics

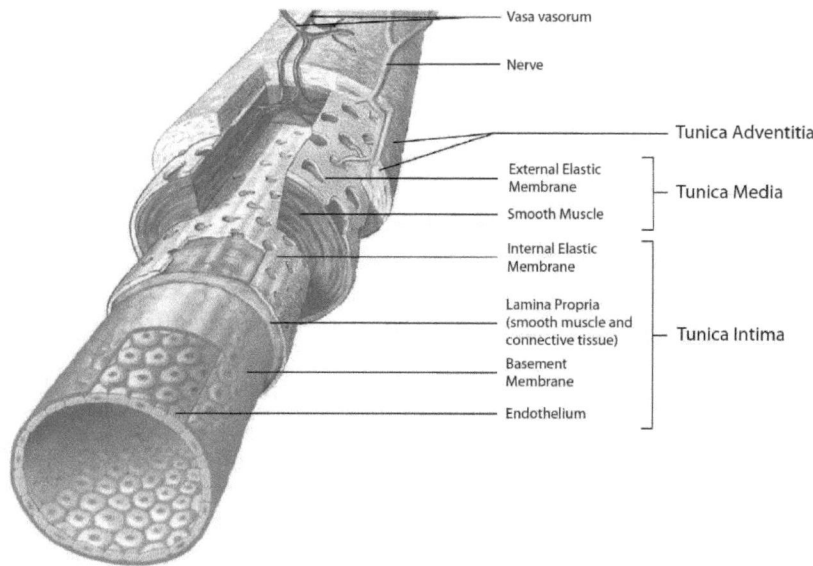

Fig. 1.1: Structural arrangement of the arterial wall, adapted from Seeley et al. [262].

an almost circumferential and coherent orientation. Woven in between the elastin sheets are bundles of tiny collagen fibrils. Proceeding from elastic to muscular arteries, the concentric layers become predominantly muscular, including up to three dozen layers of smooth muscle cells, although elastic layers also tend to be present in arteries of the arms and legs. Like elastic layers, collagenous fibers are reduced in favor of muscle.

Adventitia The adventitia makes up about 10% of the wall thickness in an elastic artery and considerably more in a muscular artery. It is surrounded by loose perivascular tissue sometimes tethering the arterial wall. The adventitial layer is essentially a dense network of collagen fibers interspersed with fibroblasts, elastic fibers, nerves, and vasa vasorum, tiny vessels providing the vessel's own blood supply. The collagen fibers in the adventitia are longitudinally oriented where individual collagen fibers show a large deviation from the mean orientation. Interestingly, the adventitia is almost absent in cerebral arteries.

Altogether, the media and adventitia provide the arterial wall with enough strength to prevent overdilatation under physiologic load. There seems to be no significant difference in the distribution of collagen in both media and adventitia of normal and atherosclerotic arteries.

1.1.2 Common Arterial Wall Diseases

Generally, homogeneous, diffuse proliferation of intimal cells is observed with age (hyperplasia). This results in an increase of extracellular matrix containing mainly dispersed collagen fibers — thus a stiffening of the wall (*arteriosclerosis*[1]).

Atherosclerosis In the common disease of *atherosclerosis* intimal components thicken and stiffen *locally* together with deposition of fatty substances, calcium, collagen fibers, cellular waste products, and fibrin. Atherosclerosis usually affects medium and large arteries and often develops near branches, bifurcations or curves and most prominent locations are the carotid bifurcation and the abdominal aorta. A complex manifestation of biomechanical and biochemical events leads to the development of lesions, known as atherosclerotic plaque. Tissue overgrowth damages the endothelium and reduces or even blocks blood flow. If this occurs in the coronary arteries, it can cause ischemic heart disease. The advanced plaque may induce thrombi which may become instable leading to rupture and emboli. This is a major cause of stroke and myocardial infarction. These diseases are considered as major cause of human mortality in the western world.

Aneurysms An aneurysm is defined as focal dilatation of the arterial wall. Its initial dilatation is nowadays regarded to be caused in part by degeneration of a portion of the wall such as medial elastin and smooth muscle cells. Aneurysms are often associated with atherosclerosis and hypertension, but traumatic injuries, chronic lung diseases, genetic disorders, gender, and smoking are risk factors as well. The two most common types are abdominal aortic aneurysms (AAA) occurring in the infrarenal aorta, and intracranial aneurysms (IA) occurring in or near the circle of Willis, the major network supplying oxygen and nutrients to the brain. The natural history of aneurysms consists of three phases: pathogenesis, enlargement, and rupture which lead to spontaneous and often lethal hemorrhage. In Fig. 1.2 two histological sections are depicted comparing a healthy and a diseased arterial wall.

Fatigue hypothesis for pathogenesis Both diseases are characterized by a structural change in the arterial wall. Thubrikar [301, 304] establishes an intriguingly simple hypothesis for the pathogenesis of atherosclerosis and aneurysms as a result of failure of the artery from fatigue in the region of stress-concentration. Some of the observed features leading to this hypothesis are

- Atherosclerotic plaques occur at branches and bifurcation where stress-concentration occurs.

- Atherosclerosis increases with increasing mean pressure and pulse pressure and is reduced by lowering mean pressure and heart rate (e.g. β-blocker treatment).

---

[1] arteriosclerosis means *any* of a group of diseases characterized by thickening and loss of elasticity of the arterial walls, where *one prominent type* is atherosclerosis (Dorland [71]).

1. Introduction to Arterial Wall Mechanics

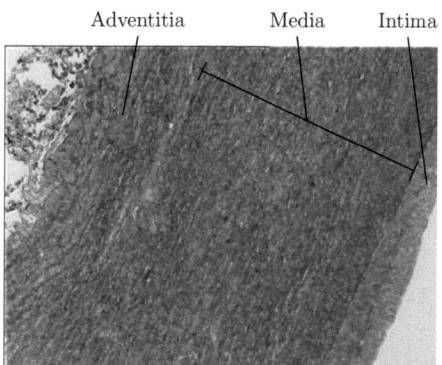
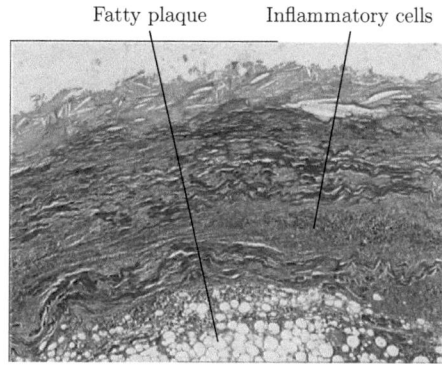

Fig. 1.2: Histological sections of healthy (left) and diseased (right) abdominal aorta, obtained from C. Reeps, Chirurgische Klinik, Gefäßchirurgie, Klinikum Rechts der Isar, Technische Universität München.

- Atherosclerosis does not occur in arteries with pressure below a certain threshold such as pulmonary arteries or veins.

The cellular mechanism is supposed to be related to a high cyclic tension of the smooth muscle cells particularly at stress-concentration regions and therefore to an injury by damaging cell-to-cell connections. This results in a stimulated SMC proliferation, promoted by penetration lipoproteins or risk factors like hypertension, diabetes, smoking, etc. On the other hand, injury to SMC may result in cell death as an initial step in the development of aneurysms. Thus, the common key factor of vascular disease is hypothesized to be the stress-induced fatigue damage of SMC and the onset of atherosclerosis or aneurysm is differentiated by the ability of their proliferation. Several convincing observations confirming this hypothesis are elaborated in Thubrikar [301]. By all means this hypothesis emphasizes the need for a comprehensive examination of in-vivo stress-state.

1.2 Structural Characteristics of the Arterial Wall

Arterial wall tissue is characterized by a macroscopic mechanical response that is highly nonlinear, incompressible, anisotropic, and (in)elastic. The arterial wall is understood as a composite structure of individual tissue components, in particular elastin and collagen, and an amorphous, gel-like, ground substance consisting mostly of water. The overall nonlinear response has been first identified by Roach and Burton [248] who isolated the contribution of elastin and collagen. They found that the response of the arterial wall to low tension is similar to that of elastin, while the response to high tension follows the characteristic stiffening of collagen. The load transfer from elastin to collagen is strain dependent and a gradual recruitment of collagen fibers results in significant stiffening at higher but still physiological blood pressure. The precise connection between elastin and collagen fibers is still unclear (Hayashi et al. [133]). The contribution of the aligned collagen fibers leads to considerable anisotropy, while the high content of water implicates incompressibility of the tissue.

Overall, arterial walls behave in an elastic manner; however several inelastic phenomena are also present. Arteries exhibit hysteresis under cyclic loading and relaxation under constant load, representing viscoelastic effects. High strain rates give stiffer response, but such strain rate dependency is not very pronounced and hysteresis stays constant across several decades of strain rate. In vitro tests of arterial tissue also show a pronounced stress softening in the first few load cycles. Similar behavior occurs in rubber ('Mullins effect') and uncoiling of the polymeric chains is thought to play a key role. After several load cycles (pre-conditioning) a repeatable stress-strain relation is observed. Further inelastic effects of damage and failure arise above the physiological loading range which may occur during medical treatments like angioplasty. In a diseased state of atherosclerosis, the involved plaques lead to significant alterations in the structural properties of the wall (Holzapfel et al. [143]).

A feature of paramount importance for characterizing the biomechanics of the arteries is the prestress in the unloaded configuration. For the definition of stress, strain, stiffness, and material symmetry, the identification of the appropriate reference configuration is essential. However, as noted by Fung [101], for soft tissue the existence of a single natural configuration is unlikely, because they grow and remodel and therefore undergo continuous irreversible structural changes. The arterial wall thus contains residual stresses which vary with location, age and disease. Two major prestress phenomena, the axial prestress and the circumferential prestress characterize the gross residual stresses within the artery. Most arteries are significantly stretched in axial direction in the basal in vivo state which is recognized by contraction of about 50% upon excision. Han and Fung [120] report in situ axial stretch in porcine and canine aortas, increasing from 1.2 near the aortic arch to 1.6 near the iliac bifurcation. Very recently, Humphrey et al. [153] pointed out the fundamental role of axial stress in compensatory arterial adaptations resulting for instance from increased blood pressure.

The second circumferential prestress effect is associated with a 'spring-open' of a radial cut arterial ring. Fung [103] and Vaishnav and Vossoughi [313] independently presented the finding that excised, intact, unloaded arterial rings open up in response to a radial cut. This implies residual stresses of bending type where the inner wall of the unloaded ring is in compression and the outer wall in tension. Light microscopy confirms this observation as the internal elastic lamina presents a certain waviness in the unloaded intact state indicating compression. It is suggested that an opening angle can serve as single measure of the residual strain, as a single radial cut captures most of the associated residual strain independently of the position of the cut (Chuong and Fung [56], Han and Fung [120]). However, as reported by Greenwald et al. [116] and Holzapfel et al. [142] pronounced differences in the pre-stretch of individual wall layers exist. For example, adventitias from aortic rings become flat, intimas open only slightly and medias spring open by more than 180°. Liu and Fung [197, 198] report detailed results on opening angles in arteries varying along the vascular tree. They compare calculations of circumferential stretch with and without taking residual strains into account resulting in significant differences.

Matsumoto and Hayashi [209] show that the opening angle increases in hypertensive vessels of rat thoracic aorta. It is assumed that such residual strains homogenize the stress and/or strain within the arterial wall at in vivo loading conditions. Delfino et al. [64] were among the first who considered residual strains at a more complex location of the carotid bifurcation by means of a finite element analysis. We will address the issue of residual strains and stresses at several points in the remainder of the present work.

1.3 Review of Computational Modeling in Vascular Mechanics

Computational technology provides powerful tools to investigate the biomechanical behavior of the vascular system (Steinmann et al. [278]). In this section a literature overview on recent computational advances improving the understanding of the healthy and diseased vascular system with emphasis on atherosclerosis and aneurysm is presented. Research in this field is based on the widely accepted assumption that cells sense and respond to their biomechanical environment which is thought to play a key role in the progression and development of arterial diseases. Furthermore, clinical intervention strategies often depend on biomechanical factors and computational techniques can also in this regard help to improve such treatments and device designs.

Research is abundant in this fast growing field of scientific inquiry, with papers ranging from clinical to experimental and from analytical to computational approaches. Below, we focus on advanced computational approaches rather than experimental and clinical studies. The reader may also consult the following review articles for further references: Steinmann et al. [278], Humphrey and Taylor [156], Vorp [320], Duraiswamy et al. [77], Holzapfel et al. [144], as well as the chapters of Taylor [293] and Holzapfel [138] in the 'Encyclopedia of Computational Mechanics'.

1.3.1 Patient-Specific Modeling

Progress in medical imaging technology, such as computed tomography (CT) and magnetic resonance imaging (MRI), sophisticated image processing techniques and high-performance computing allowed simulation of physiologically pulsatile flow patterns in anatomically realistic or even patient-specific geometries. The importance of taking such image-based geometries into account to determine the local environment was soon established (Steinman [279]). Today, noninvasive in vivo methods include CT, ultrasound and echocardiography, and MRI. CT is generally used to evaluate anatomy, while ultrasound and echocardiography can be used to acquire the in-plane component of velocity. MRI is a unique imaging technology that can acquire both three-dimensional anatomy and velocity fields throughout the cardiac cycle. Unlike CT, which is limited to axial slice acquisitions, planes and volumes of arbitrary orientation can be obtained. Details on the physics of MRI can be found at Taylor and Draney [294] and references therein.

Image processing allows segmentation of the region of interest and subsequently a three-dimensional geometric representation of the considered region can be generated. If automatic segmentation fails due to lack of contrast time-consuming manual segmentation becomes unavoidable. Tools and algorithms for processing, editing and smoothing of the data are widespread and high-quality geometries are available today. However, specific regions such as the lung or the arterial wall still pose high demands on image technology and a lot of research is in progress in this field. Additional information such as the microstructure of the considered tissue would provide important advancement for vascular mechanics. For blood flow simulations realistic inflow boundary conditions from MRI data would improve the quality and accuracy of the flow simulations.

1.3.2 Computational Hemodynamics

Hemodynamics refers to the physiology dealing with the forces involved in the circulation of the blood. It is known to provide stimuli for many acute and chronic biologic adaptations and local hemodynamic factors including flow rate, (wall) shear stress, and pressure forces triggering changes in cardiac output and downstream vascular resistance. Changes in blood velocity and pressure fields, sensed at cellular level in the endothelium, initiate a cascade of biochemical signals leading to hierarchical reorganization across molecular, cellular, tissue, and system scales. Because it is observed that atherosclerosis is localized at branches and bends of the arterial tree and at these locations complex flow conditions are present it is hypothesized that such complex flow may be associated with the onset of atherosclerosis (Caro et al. [53], Zarins et al. [338]). The complex flow is characterized by stagnation and recirculation which result in extremal wall shear stresses. Similarly, the localization of aneurysmal disease is hypothesized to be influenced by hemodynamic conditions like flow stagnation and pressure wave amplification. For cerebral aneurysms there are experimental and clinical findings that confirm this conjecture, see Meng et al. [211] and Kondo et al. [174]. Additionally, since reconstructive surgeries or catheter-based interventions alter hemodynamic conditions, methods to model blood flow have increasing application in clinical decision-making and surgery. Hence, the biomechanics community has had long-standing interest in understanding hemodynamic patterns, and computational methods have been widely applied to the quantification of such factors in relation to the genesis, progression, and clinical consequences of vascular disease.

Since the pioneering work of Perktold and coworkers [230, 229] modeling a carotid bifurcation in the late 1980s, computational methods have been used extensively to solve such problems. The governing equations for three-dimensional blood flow in large vessels under the assumption of an incompressible homogenous, Newtonian fluid flow in a fixed domain consist of the Navier-Stokes equations and suitable initial and boundary conditions. Taylor, Hughes and Zarins [296, 297] in the late 1990s significantly contributed to assess hemodynamic features by advanced computational modeling of three-dimensional physiologically realistic problems.

The progress in computational methods also entailed extensive work on the optimization of surgical procedure designs. For instance, bypass surgery was investigated by a number of researchers and design and shape of the joining graft with the artery could be improved with respect to hemodynamical factors, see Lei et al. [193], among others. With the construction of patient-specific geometries and efficient computational fluid dynamic (CFD) strategies it was possible to predict changes in blood flow resulting from therapeutic interventions for individual patients. The group of Taylor and colleagues [295, 254] are leading proponents of this 'predictive medicine' approach (Steinman et al. [278]) and have recently driven this concept to an automated process.

However, more recently there emerged an increasing agreement that hemodynamics alone could not fully explain the onset and development of such diseases. For instance, Steinman et al. state from their CFD study [280] that it "failed to find a significant relationship between wall thickness and wall shear stress variables when considering data from the whole carotid bifurcation, hinting at a more complex relationship between local hemodynamic factors and atherosclerosis that will no doubt be the subject of further scrutiny with these novel techniques."[278] When comparing various computed fluid dynamic variables with histological markers of atherosclerosis, Kaazempur-Mofrad et al. [162] found only inconclusive correlations. Thubrikar [301] argues that despite the voluminous research, basic observations such as the effect of hypertension or β-blocker treatment on atherosclerosis remain unexplained by considering only hemodynamics. With respect to enlargement or rupture of intracranial aneurysms, also Humphrey and Taylor [156] state in one of their most recent papers that there has been no consensus as to which hemodynamic factors are important. Furthermore, they conclude that although the move to image-based, patient-specific models has yielded more realistic flow patterns, there has been little progress in measuring and then assigning physiologically realistic inlet flow waveforms or outlet boundary conditions. Therefore, several researchers overcame these shortcomings by abandoning the assumption of rigid walls and taking fluid-structure-interaction (FSI) into account.

1.3.3 Fluid-Structure-Interaction Models

The assumption of a rigid wall in blood flow simulations may give insight into the blood velocity field. However for the pressure field, the assumption of rigid vessel walls precludes wave propagation phenomena and thus results change fundamentally. The difficulty of solving the coupled blood flow-vessel deformation problem (Quarteroni et al. [237]) often prohibits these type of simulations and pure blood flow results are still popular, see for example Antiga et al. [8]. But the progress in fluid-structure-interaction methods, algorithmic performance and computing power enabled such simulations. A comprehensive examination of FSI with finite elements can be found in Wall [324].

While initially simplified or reduced geometries were considered, for instance in Perktold and Rappitsch [231], or van de Vosse et al. [315], more recently methods for three-dimensional FSI-simulations on realistic or patient-specific geometries have been presented by a few research groups, for example Le Tallec, Gerbeau and colleagues [191, 112, 113], Leuprecht, Perktold, and colleagues [195, 194], by Heil [134], the group of Torii, Oshima, Kobayashi, Takagi, and Tezduyar [307, 306, 308], and the group of Taylor and colleagues [86].

An important issue in FSI simulations as well as in pure fluid simulations is the application of correct boundary conditions. Special considerations are necessary for inlet and outlet boundary conditions for both velocity and pressure fields in the fluid and tractions on the vessel wall. Formaggia et al. [94] and Vignon-Clementel et al. [318] considerably contributed to this topic. Another issue is the fluid–solid coupling scheme, where due to the large structure deformation and typically similar densities between solid and fluid only strong coupling schemes seem suitable, often applied within an monolithic approach (see Küttler et al. [183]).

However, computational solutions of FSI problems in biomechanical applications are still challenging in many aspects of the participating solution algorithms. A promising new solution approach — 'isogeometric analysis' — has been presented by Hughes et al. [148]. The potential of the involved NURBS geometry approximation is beneficial in complex geometries, but has its merits also as solution approximation. We have explored the capabilities in the field of structural optimization, see Frenzel et al. [100] and Wall et al. [326]. Isogeometric analysis is applied for blood flow FSI simulations by Bazilevs et al. [24] and with emphasis on patient-specific models by Zhang et al. [339]. Calo et al. [51] have extended the isogeometric bloodflow FSI framework to drug transport in arteries, however restricted to small-strain/small-deformation. Further details on mass-transfer problems can be found at Kaazempur-Mofrad and Ethier [161], or Comerford and David [57].

Regarding the onset and development of AAA a couple of studies have been performed with FSI simulations. Di Martino et al. [66] were probably the first who considered a realistic geometry of the aneurysm. However, their simulation was limited to a small-strain/small-deformation structural model. Wolters et al. [332] present large-strain/large-deformation model together with a methodology to generate good quality FSI meshes. A shortcoming of these simulations are the rather simple material laws for the modeling of the wall (see below). Another limiting factor is the widely-used assumption that the wall thickness is constant, which significantly influences wall stress results in FSI simulations, see Scotti et al. [261]. With respect to surgical procedures, Li and Kleinstreuer [196, 168] have dealt with FSI simulations in idealized stented AAA geometries determining the influence of endovascular grafts.

In summary, FSI models of the vascular system deepen the insight into significant local hemodynamic characteristics which in turn influence the structural load impact on the wall. However, no direct correlation of such simulations with disease has been reported yet. From a computational modeling point of view, the applied structural wall models are still quite simplified and therefore limited in their insight into the realistic wall stress. Steinmann et al. [278] argue that flow dynamics resulting from an FSI simulation of an anastomosis technique performed by Leuprecht et al. [195] plays only a minor role in bed hyperplasia, whereas structural analysis of the wall revealed a much higher relationship. Humphrey and Taylor conclude that "although the move to image-based, patient-specific models has yielded more realistic velocity and shear stress fields compared with idealized models, there has been little progress in measuring and then assigning physiologically realistic inlet flow waveforms, outlet boundary conditions, or wall properties."[156] They stress that a sophisticated arterial wall modeling needs to be integrated into future FSI approaches.

1.3.4 Advanced Structural Wall Models

The characteristics of the arterial wall, as described in Section 1.2, are quite extensive and structural wall models generally account for only a subset of them. A comprehensive constitutive model probably remains an elusive goal, but significant progress has been achieved in recent years and the complexity of lately developed models is remarkable. In the following, we review some major contributions focusing on three-dimensional continuum-based approaches suited for implementation in computational methods, in our case the finite element method. Comprehensive reviews of mechanical models for arteries are presented by Vito and Dixon [319] and Kalita and Schaefer [164].

Parallel to the research in hemodynamics, structural modeling of arteries has been a topic of tremendous interest. Unquestionably, the prominent researcher in the field of continuum based structural arterial wall modeling is Yuan Cheng Fung, who is known as the "father of biomechanics" (Kalita and Schaefer [164]). His classical model of 1979 is often used as basis for further developments of constitutive equations, , see Fung *et al.* [106]. It rests upon a thorough non-linear continuum mechanical basis defined in terms of a strain-energy function and has been generalized by Chuong and Fung in 1983 [55]. The applied exponential form of the strain-energy function, suggested already 1972 by Demiray [65] accounts for the stiffening of soft tissue in large strains, whereas Takamizawa and Hayashi [291] propose a logarithmic form. Specified for an idealized cylindrical artery orthotropic response in circumferential, radial and axial direction is modeled, however the formulation is based on the corresponding circumferential, radial and axial strains. Holzapfel *et al.* [146] have adopted the exponential form of Fung and added an isotropic Neo-Hookean part and with this function the characteristic 'S-shaped' stress-strain relationship was captured better. Subsequently, Holzapfel *et al.* [140] reformulated the Fung-type orthotropic part in terms of structural invariants leading to a general anisotropic model. Thus, the stability issues of the Fung-model were overcome, and the Holzapfel-model was shown to be stable, see Ogden *et al.* [221]. Another benefit is the physiological relevance as the anisotropic part can be identified with reinforcing (collagen) fiber contributions. Such a '(micro-)structural model' is usually preferred because it facilitates parameter identification and interpretation of results. This model has gained high popularity in recent years and several extensions, for instance viscoelasticity, fiber dispersion, fiber remodeling have been implemented which is discussed in detail in the remainder of the present work.

While the previous models have considered only the passive response of the artery, Rachev and Hayashi [239] model vascular smooth muscle contraction. Moreover, Humphrey and Na [154] include blood-flow-induced shear stress, dynamic circumferential wall motion, smooth muscle activation, perivascular tethering, wall heterogeneity, and geometric nonlinearities in a cylindrically orthotropic, residually stresses model, resulting in ten unknown parameters.

According to Vito and Dixon [319], this model is the most comprehensive elastic model, but due to limitations of current experimental methods validation is difficult.

However, complex three-dimensional models are also sensitive to their defining parameters as well as to the influence of geometry, boundary conditions, etc. Furthermore, the involved parameters are often difficult to obtain experimentally, especially in vivo conditions. Therefore, the typically necessary fitting of the models to some measurement data is susceptible to mistakes. A classical quote of Enrico Fermi by Freeman Dyson [79] reflects this *law of parsimony*: "I remember my friend Johnny von Neumann used to say, with four parameters I can fit an elephant, and with five I can make him wiggle his trunk." One should bear in mind that also a number of simpler models exist which are fairly accurate and remain popular for global considerations, see Kalita and Schaefer [164] for details.

More recently, constitutive models with phenomenologically motivated microstructural considerations based on statistical mechanics of chain molecules such as collagen fibers became popular. Homogenization from the molecular microscale to the macroscale of the tissue is performed via the concept of chain network models. A chain network consists of a representative unit-cell, for example the eight-chain cell proposed by Arruda and Boyce [10]. Bischoff *et al.* [32, 33] have extended this model to orthotropy and a first application for pulmonary arteries is presented by Zhang *et al.* [340]. The benefit is the limited number of parameters which have a clear physical interpretation.

With respect to vascular diseases the research in arterial wall mechanics was intensively driven by the topic of aneurysms, especially abdominal aortic aneurysms. The obvious premise is that AAA rupture follows the principle of material failure, when the mural stress exceeds the strength of the wall. The quantification of a rupture risk has been based on the law of Laplace leading to a 'maximal diameter criterion'. However this seems to be an oversimplification due to the complex geometry of aneurysms. Vorp, Raghavan, Vande Geest, Di Martino, Wang, and colleagues have significantly contributed to this field of research and especially the corresponding work of Fillinger *et al.* [88, 87] was recognized also in the medical community. Further studies were presented, among others, by Thubrikar [302, 303]. We refer to the comprehensive review of Vorp [320] for details and further references. Usually, these researchers performed nonlinear finite element analyses, employing patient-specific geometries and treated the aneurysmal wall as nonlinearly elastic. Also the role of the intraluminal thrombus was studied, see for instance Vorp *et al.* [321], Wang *et al.* [327], Dam *et al.* [314], Vande Geest and colleagues [316, 12]. However, shortcomings of these simulations are assumptions of isotropic, homogeneous and uniformly thick walls without considering prestress.

Regarding complex arterial wall constitutive laws together with atherosclerosis, plaque rupture was modeled by Ohayon *et al.* [222] (see also Richardson [247]), but research is mainly focused on clinical treatments of angioplasty and stenting, instead of the onset and development of the disease itself. With respect to angioplasty Holzapfel *et al.* [145] present an advanced structural model obtained from MRI. With respect to stenting, computational methods have shown to be a useful tool for the design of stents in studies of free expansion and the interaction of stents with the balloon and the arterial wall. Among others, Holzapfel *et al.* [144], the group of Migliavacca [216], De Beule, Verdonck and colleagues [63] and Wu, Wang and colleagues [336] deal with stenting.

All the previously discussed approaches have focused to describe the arterial wall response at a particular instant, not taking into account the development of the tissue due to remodeling. Topics of growth, remodeling and adaptation to perturbed loading have gained interest in recent years. Initiated by the work of Taber [288, 289] and Rachev and colleagues [240, 238] mathematical models have been developed and diverse manifestations of vessel growth can be predicted by such models (Humphrey and Taylor [156]). However, these models are limited to a uniform material ignoring structural constituents like fibers, muscle cells, etc. If the fiber structure is accounted for in terms of anisotropy, remodeling should be differentiated into morphogenesis, growth and fiber remodeling/reorientation (Taber [288]). The underlying hypothesis is always that the tissue adapts to its environment seeking for a kind of optimal configuration. In a constrained mixture theory, Humphrey and Rajagopal [155] have taken evolving properties and turnover rates of individual constituents like fibers and smooth muscles into account. Very recently, Humphrey and Taylor [156] have proposed to couple this approach with a FSI simulation of patient-specific vascular problems, see also Figueroa *et al.* [85]. Including remodeling approaches drives complexity of arterial wall modeling into a new stage, and research and application of such models is still in its infancy.

1.4 Objectives of the Present Work

The present work is dedicated to the investigation of the stress field in patient-specific arterial wall geometries by means of the finite element method. As detailed above, analyses of the arterial wall stress state require advanced structural methods due to the involved nonlinearities stemming from, among others, large deformations, large strains, anisotropy, and viscoelasticity. In addition, such analyses are inevitably three-dimensional, and the involved patient-specific geometries are complex, necessitating even more sophisticated solution methods. At the same time, the computational methodology needs to perform the analyses with an acceptable effort in order to enable future studies which correlate clinical data with simulation data to obtain further insight into pathologies and medical treatment.

Such highly complex computational demands are far beyond the typical capabilities of commercially available software packages. Therefore, a major goal of the present work was to develop a computational toolchain to investigate the stress field in patient-specific arterial wall geometries by means of the finite element method. The toolchain incorporates medical image acquisition, segmentation, mesh generation, simulation, and postprocessing, where we expand upon available software, focusing on the generation of the arterial wall model and the structural simulation methods. In addition, the toolchain is implemented into the in-house research code **baci** which is capable of solving coupled FSI- and mass-transfer simulations. Therefore this approach is particularly promising with regard to intriguing future research questions, such as the simulation of fully coupled hemodynamic, mass-transfer, and advanced structural simulations.

We cover three major topics with respect to finite element modeling of the structural arterial wall, namely (1) the three-dimensional finite element technology, (2) constitutive laws suited for modeling of the arterial wall and (3) the issue of fiber remodeling. After a short review of continuum mechanical preliminaries in **Chapter 2**, we elaborate upon the first topic of three-dimensional finite element technology in **Chapter 3**. It is well known that low order finite elements suffer from so-called locking defects. To accurately model the arterial wall with such elements, we propose specific advanced element technology to overcome these problems. Since the arterial wall is usually regarded as a thin-walled structure, special so-called solid-shell elements are beneficial to enhance computational efficiency.

In response to the shape peculiarities of patient-specific geometries, we derive hexahedral and wedge-shaped solid-elements. Within these solid-shell element formulations, the issue of incompressibility, typically present in biomechanical applications, needs to be tackled, and several further locking defects in bending need to be eliminated. Popular computational approaches in biomechanics seem insufficient in addressing these problems. For example, the widely applied mixed Jacobian-pressure formulation proposed by Simo *et al.* [275] is not suited to efficiently model thin-walled structures. To the authors' knowledge, the interplay between finite element technology methods and biomechanical problems involving incompressibility has not been addressed in the literature so far. Finally, the general performance of the proposed three-dimensional elements specifically designed to satisfy our needs in arterial wall modeling is evaluated in popular benchmark examples.

In **Chapter 4**, we address the second topic, the constitutive laws suited for modeling of the arterial wall. As described above, the arterial wall has several demanding characteristics which need to be captured. Within this chapter a number of recent constitutive laws considering anisotropy, viscoelasticity, and the underlying microstructure are presented. This includes rather simple isotropic laws, popular anisotropic laws based on structural tensors, and a microstructurally motivated anisotropic continuum-chain-network model. Moreover, the extension to viscoelasticity is discussed. All models are implemented into the in-house finite element code and a set of numerical examples demonstrates the different model features.

Chapter 5 deals with the third topic of fiber remodeling. As mentioned above, we refer to the term remodeling as describing the microstructural change of the tissue in response to its — in our case purely mechanical — environment. After a literature review and a concise problem definition, we focus on the alignment of the fiber pattern. Therefore, recent approaches are discussed in detail and two strategies are implemented. We apply these remodeling strategies to geometries with increasing complexity, ranging from idealized arteries and tendons to more complex bifurcation geometries. Strengths and weaknesses of the two approaches are evaluated and further advancements proposed. The presented strategies are generally capable of reproducing the histologically observed, physiological fiber pattern of an idealized arterial wall section. For more complex geometries the remodeling strategy yields insight into a mechanically reasonable fiber pattern. To this end, we assess whether resulting fiber patterns are physiologically reasonable and correlate our findings with histological data reported in the literature. However, a limiting factor of these methods for patient-specific simulations is the fact, that the boundary conditions of the physiological environment remain unknown and need to be further investigated.

In order to efficiently simulate patient-specific geometries, additional steps are necessary, including the generation of detailed three-dimensional models of the arterial wall geometry, and an adequate alignment of the fibers even in complex geometries. These steps are described in **Chapter 6**. Modern medical image technology provides high resolution segmentation data generating detailed three-dimensional models of the arterial vessel lumen. However, the arterial wall itself is in fact not accessible via segmentation. We therefore develop an algorithm which generates the wall by extrusion of the meshed vessel model. We propose to use further information provided by the computed vessel centerline to enhance the quality of the wall model. For instance, the wall thickness can be based on a fitted lumen diameter. In addition, we suggest aligning the fiber pattern to a local coordinate system obtained from the centerline which enables the fibers to follow the winding vessel even in complex geometries.

Finally, in **Chapter 6**, we also present structural simulations of two patient-specific arterial wall geometries, one of the aortic arch and one of the iliac bifurcations, and explore the effect of the methodologies proposed throughout the work on these examples. A conclusion of the presented work together with an outlook on future research are given in **Chapter 7**.

2. Some Nonlinear Continuum Mechanics

The present chapter shortly introduces the necessary fundamentals of continuum mechanics. Comprehensive elaboration of this topic can be found in the abundant literature. Note that sometimes a detailed discussion of specific continuum mechanical questions is shifted to subsequent chapters where they are in close relationship to the corresponding topic.

2.1 Kinematics

We introduce the kinematics to describe the motion and deformation of a homogenous body which is seen as a continuous compact set of material points or particles. We define $\mathcal{B}_0 \subset \mathbb{R}^3$ as the reference configuration of the material body at time $t = t_0$ and $\mathcal{B}_t \subset \mathbb{R}^3$ as its current configuration. The body transforms from its reference configuration (material frame) to its current configuration (spatial frame) and the corresponding nonlinear map

$$\varphi_t : \mathcal{B}_0 \to \mathcal{B}_t \tag{2.1}$$

is required to be unique and continuously differentiable. At a fixed time $t \in \mathbb{R}_+$ material points $\mathbf{X} \in \mathcal{B}_0$ of the reference configuration are mapped to points $\mathbf{x} \in \mathcal{B}_t$ of the current configuration

$$\varphi_t : \mathbf{X} \mapsto \mathbf{x} = \varphi_t(\mathbf{X}, t) = \mathbf{x}(\mathbf{X}, t). \tag{2.2}$$

The inverse map is thus uniquely defined as $\mathbf{X} = \varphi_t^{-1}(\mathbf{x}, t)$ at every time t. The difference between current and reference configuration is the deformation and thus the deformation vector is given as

$$\mathbf{u}(\mathbf{X}, t) = \mathbf{x}(\mathbf{X}, t) - \mathbf{X}. \tag{2.3}$$

The temporal change of material points is defined as velocity field in the material and spatial frame

$$\mathbf{V}(\mathbf{X}, t) = \frac{\mathrm{d}\varphi_t(\mathbf{X}, t)}{\mathrm{d}t} = \dot{\varphi}_t(\mathbf{X}, t) = \dot{\mathbf{x}}(\mathbf{X}, t) \tag{2.4}$$

$$\mathbf{v}(\mathbf{x}, t) = \mathbf{V}(\varphi_t^{-1}(\mathbf{x}, t), t) \tag{2.5}$$

and furthermore material time differentiation yields the material acceleration field as

$$\mathbf{A}(\mathbf{X}, t) = \frac{\mathrm{d}\mathbf{V}(\mathbf{X}, t)}{\mathrm{d}t} = \dot{\mathbf{V}}(\mathbf{X}, t) = \ddot{\mathbf{x}}(\mathbf{X}, t). \tag{2.6}$$

The spatial acceleration field is obtained as

$$a(\mathbf{x}, t) = \mathbf{A}(\varphi_t^{-1}(\mathbf{x}, t), t)$$
$$= \frac{d}{dt}\mathbf{v}(\mathbf{x}, t) = \frac{\partial \mathbf{v}}{\partial \mathbf{x}} \cdot \frac{\partial \mathbf{x}}{\partial t} + \frac{\partial \mathbf{v}}{\partial t} = \operatorname{grad} \mathbf{v} \cdot \mathbf{v} + \frac{\partial \mathbf{v}}{\partial t} \qquad (2.7)$$

where the first is the convective term and the second is the local or partial time derivative.

The material deformation gradient

$$\boldsymbol{F}(\mathbf{X}) := \operatorname{Grad} \mathbf{x} = \frac{\partial \mathbf{x}}{\partial \mathbf{X}} \qquad (2.8)$$

is an essential kinematic quantity defined as the partial derivative of the nonlinear deformation map $\mathbf{x} = \varphi_t(\mathbf{X}, t)$ with respect to the coordinates $\mathbf{X}$. Being a two-field tensor, it maps an infinitesimal line element $d\mathbf{X}$ at position $\mathbf{X}$ of the reference configuration to the infinitesimal line element $d\mathbf{x}$ at $\mathbf{x}$ of the current configuration.

Uniqueness of the mapping φ_t requires $\boldsymbol{F}$ not to be singular and the inverse

$$\boldsymbol{F}^{-1} = \operatorname{grad} \mathbf{X} = \frac{\partial \mathbf{X}}{\partial \mathbf{x}} \qquad (2.9)$$

is well-defined. This is equivalent with $J := \det \boldsymbol{F} \neq 0$ where we introduce the Jacobi determinant J. The requested smoothness of $\varphi_t(\mathbf{X}, t)$ implies

$$J = \det \boldsymbol{F} > 0. \qquad (2.10)$$

Furthermore, transformations of the infinitesimal line, area and volume elements read

$$d\mathbf{x} = \boldsymbol{F} d\mathbf{X} \qquad (2.11)$$
$$d\mathbf{a} = J\boldsymbol{F}^{-T} d\mathbf{A} = \operatorname{cof} \boldsymbol{F} \ d\mathbf{A} \qquad (2.12)$$
$$dv = J dV \qquad (2.13)$$

and together with (2.10) the physically intuitive argument that a body must not contain negative volume elements and must not penetrate itself is satisfied.

Though the deformation gradient describes the deformation uniquely it still contains rigid body motions and is thus not suitable to describe the strain of a body. However, every deformation may be split into rigid body motion and stretch as

$$\boldsymbol{F} = \boldsymbol{R}\boldsymbol{U} = \boldsymbol{v}\boldsymbol{R} \qquad (2.14)$$

where $\boldsymbol{R}$ is an orthogonal rotation tensor with $\boldsymbol{R}^{-1} = \boldsymbol{R}^T$ and the symmetric positive definite tensors $\boldsymbol{U}$ and $\boldsymbol{v}$ define the material and spatial stretch tensors, referred to as right and left stretch tensors, respectively.

We introduce the right Cauchy-Green tensor $\boldsymbol{C}$ and the left Cauchy-Green or Finger tensor $\boldsymbol{b}$ as important strain quantities in continuum mechanics

$$\boldsymbol{C} := \boldsymbol{F}^T \boldsymbol{F} \quad = (\boldsymbol{R}\boldsymbol{U})^T \boldsymbol{R}\boldsymbol{U} \quad = \boldsymbol{U}^T \boldsymbol{R}^T \boldsymbol{R}\boldsymbol{U} \quad = \boldsymbol{U}^T \boldsymbol{U} \quad = \boldsymbol{U}^2 \qquad (2.15)$$
$$\boldsymbol{b} := \boldsymbol{F}\boldsymbol{F}^T \quad = \boldsymbol{v}\boldsymbol{R}(\boldsymbol{v}\boldsymbol{R})^T \quad = \boldsymbol{v}\boldsymbol{R}\boldsymbol{R}^T \boldsymbol{v}^T \quad = \boldsymbol{v}\boldsymbol{v}^T \quad = \boldsymbol{v}^2. \qquad (2.16)$$

2. Some Nonlinear Continuum Mechanics

Another quantitative measure for strain is the difference between the squares of infinitesimal line elements in current and reference configuration

$$s = \mathrm{d}\mathbf{x} \cdot \mathrm{d}\mathbf{x} - \mathrm{d}\mathbf{X} \cdot \mathrm{d}\mathbf{X} \tag{2.17}$$

and we may define the Green-Lagrangean strain tensor $\mathbf{E}$ and the Euler-Almansi strain tensor $\mathbf{e}$

$$\mathbf{E} := \frac{1}{2}(\mathbf{C} - \mathbf{G}) \tag{2.18}$$

$$\mathbf{e} := \frac{1}{2}(\mathbf{g} - \mathbf{b}^{-1}) \tag{2.19}$$

where $\mathbf{G}$ and $\mathbf{g}$ are the second order covariant metric tensors. As we restrict ourselves to a uniform, fixed, Cartesian coordinate frame they both reduce to identity, $\mathbf{G} = \mathbf{g} = \mathbf{I}$.

As the right stretch tensor $\mathbf{U}$ is symmetric and positive definite the spectral decomposition yields three real positive eigenvalues λ_i and the corresponding real eigenvectors $\mathbf{N}_i$. With $\mathbf{N}_i \cdot \mathbf{N}_j = \delta_{ij}$ as orthonormal basis the spectral decomposition is given as

$$\mathbf{U} = \sum_{i=1}^{3} \lambda_i \, \mathbf{N}_i \otimes \mathbf{N}_i \tag{2.20}$$

and the eigenvalues λ_i are called principal stretches. They represent the quotient of deformed versus undeformed length in principal directions. It yields for the right Cauchy-Green tensor

$$\mathbf{C} = \sum_{i=1}^{3} \lambda_i^2 \, \mathbf{N}_i \otimes \mathbf{N}_i. \tag{2.21}$$

The left stretch tensor and the left Cauchy-Green tensor are composed of the same corresponding eigenvalues, only the eigenvectors as orthonormal basis change yielding

$$\mathbf{b} = \sum_{i=1}^{3} \lambda_i^2 \, \mathbf{n}_i \otimes \mathbf{n}_i \quad \text{and} \quad \mathbf{v} = \sum_{i=1}^{3} \lambda_i \, \mathbf{n}_i \otimes \mathbf{n}_i. \tag{2.22}$$

Finally, we introduce some strain rate forms, such as the material velocity gradient

$$\mathbf{L} := \dot{\mathbf{F}} = \frac{\mathrm{d}}{\mathrm{d}t}\left(\frac{\partial \boldsymbol{\varphi}}{\partial \mathbf{X}}\right) = \frac{\partial}{\partial \mathbf{X}}\left(\frac{\mathrm{d}\boldsymbol{\varphi}}{\mathrm{d}t}\right) = \frac{\partial \mathbf{V}}{\partial \mathbf{X}} = \mathrm{Grad}\,\mathbf{V}, \tag{2.23}$$

the material strain rate

$$\dot{\mathbf{E}} = \frac{\mathrm{d}}{\mathrm{d}t}\left(\frac{1}{2}(\mathbf{F}^{\mathrm{T}}\mathbf{F} - \mathbf{I})\right) = \frac{1}{2}(\dot{\mathbf{F}}^{\mathrm{T}}\mathbf{F} + \mathbf{F}^{\mathrm{T}}\dot{\mathbf{F}}) = \frac{1}{2}\dot{\mathbf{C}}, \tag{2.24}$$

and the spatial velocity gradient

$$\mathbf{l} := \dot{\mathbf{F}}\mathbf{F}^{-1} = \frac{\partial \mathbf{v}}{\partial \mathbf{x}} = \mathrm{grad}\,\mathbf{v} \tag{2.25}$$

which results from a *push-forward* operation of $\mathbf{L}$. For the definition of *push-foward* and *pull-back* of tensor quantities we refer to the literature specified above. The spatial velocity gradient can be separated into symmetric and skew symmetric parts $\mathbf{l} = \mathbf{d} + \mathbf{w}$ with

$$\mathbf{d} := \frac{1}{2}(\mathbf{l} + \mathbf{l}^{\mathrm{T}}) = \mathrm{sym}\,\mathbf{l} \quad \text{and} \quad \mathbf{w} := \frac{1}{2}(\mathbf{l} - \mathbf{l}^{\mathrm{T}}) = \mathrm{skew}\,\mathbf{l} \tag{2.26}$$

where d is called spatial strain velocity gradient and w is referred to as spatial spin tensor.

Derivatives of infinitesimal line, area, and volume elements with respect to time are obtained as

$$\mathrm{d}\dot{\mathbf{x}} = \boldsymbol{l}\mathrm{d}\boldsymbol{x}, \qquad \mathrm{d}\dot{\mathbf{a}} = \operatorname{div}\mathbf{v}\,\mathrm{d}\mathbf{a} - \boldsymbol{l}^{\mathrm{T}}\mathrm{d}\mathbf{a}, \qquad \mathrm{d}\dot{v} = \operatorname{div}\mathbf{v}\,\mathrm{d}v \qquad (2.27)$$

and for the Jacobi-determinant we obtain

$$\dot{J} = \frac{\partial \det \boldsymbol{F}}{\partial \boldsymbol{F}} : \frac{\partial \boldsymbol{F}}{\partial t} = J\boldsymbol{F}^{-\mathrm{T}} : \dot{\boldsymbol{F}} = J\operatorname{tr}\boldsymbol{l} = J\operatorname{div}\mathbf{v}. \qquad (2.28)$$

2.2 Stress Concept

As consequence of motion and deformation of interacting bodies there exists *stress* between material points within the body. We most generally introduce the quotient

$$\mathbf{t} = \frac{\mathrm{d}\mathbf{f}_a}{\mathrm{d}a} \qquad (2.29)$$

as stress vector relating the force resultant $\mathrm{d}\mathbf{f}_a$ to the infinitesimal area $\mathrm{d}a$. The orientation of the area element is represented by its spatial normal $\mathbf{n}$. The Cauchy theorem

$$\mathbf{t} = \boldsymbol{\sigma} \cdot \mathbf{n} \qquad (2.30)$$

results from equilibrium condition at the infinitesimal tetrahedral element and maps the normal vector to the to the resulting force vector. The *Cauchy* stresses $\boldsymbol{\sigma}$ reflect the real internal stress state within a body at its current configuration. Using transformation (2.12) we get

$$\mathrm{d}\mathbf{f}_a = \boldsymbol{P} \cdot \mathrm{d}\mathbf{A}, \qquad \text{introducing} \qquad \boldsymbol{P} = J\boldsymbol{\sigma}\boldsymbol{F}^{-\mathrm{T}} \qquad (2.31)$$

as *first Piola-Kirchhoff* stress tensor. It maps the material area element $\mathrm{d}\mathbf{A}$ onto the spatial force resultant $\mathrm{d}\mathbf{f}_a$ and is thus a two-field tensor. We apply a *pull-back* to the spatial force resultant to introduce the fictitious force vector $\mathrm{d}\boldsymbol{F}_a = \boldsymbol{F}^{-1} \cdot \mathrm{d}\mathbf{f}_a$ and therefby define

$$\mathrm{d}\boldsymbol{F}_a = \boldsymbol{S} \cdot \mathrm{d}\mathbf{A}, \qquad \text{referring to} \qquad \boldsymbol{S} = \boldsymbol{F}^{-1}\boldsymbol{P} = J\boldsymbol{F}^{-1}\boldsymbol{\sigma}\boldsymbol{F}^{-\mathrm{T}} \qquad (2.32)$$

as *second Piola-Kirchhoff* stress tensor which is completely related to the material frame. Another stress tensor is the *Kirchhoff* stress defined as

$$\boldsymbol{\tau} := J\boldsymbol{\sigma} = \boldsymbol{F}\boldsymbol{S}\boldsymbol{F}^{\mathrm{T}} \qquad (2.33)$$

We may also introduce the additive split

$$\boldsymbol{\sigma} = \operatorname{dev}\boldsymbol{\sigma} + p\boldsymbol{I}, \qquad \text{with} \qquad p = \frac{1}{3}\operatorname{tr}\boldsymbol{\sigma} \qquad \text{and} \qquad \operatorname{dev}\boldsymbol{\sigma} = \boldsymbol{\sigma} - p\boldsymbol{I} \qquad (2.34)$$

of the Cauchy stress into a deviatoric stress part and a hydrostatic pressure part. This split is automatically obtained if an isochoric-volumetric split of the underlying strain energy function is applied, as discussed in detail in Section 4.1.3.

2.3 Balance Principles and Entropy

Conservation of mass requires that the total amount of mass of a body keeps constant. With the density $\rho = \rho(\mathbf{x}, t)$ related to the spatial volume element dv and the density $\rho_0 = \rho_0(\mathbf{X})$ related to the material volume element dV it holds

$$m = \int_{\mathcal{B}_t} \rho \, dv = \int_{\mathcal{B}_0} \rho_0 \, dV = \text{const} \qquad (2.35)$$

with $\rho_0 = J\rho$. The material (total) time derivative must vanish and consideration of an arbitrary portion of the body yields the local form in spatial and material frame

$$\dot{\rho} + \rho \operatorname{div} \dot{\mathbf{x}} = 0 \quad \text{and} \quad \dot{\rho}_0 = 0. \qquad (2.36)$$

Balance of linear momentum states that the material time derivative of linear momentum equals the sum of all applied volume and surface forces:

$$\frac{d}{dt} \int_{\mathcal{B}_t} \rho \dot{\mathbf{x}} \, dv = \int_{\mathcal{B}_t} \rho \mathbf{b}_0 \, dv + \int_{\partial \mathcal{B}_t} \mathbf{t} \, da \qquad (2.37)$$

where $\partial \mathcal{B}_t$ is boundary of the body at current configuration and $\mathbf{b}_0$ are external volume forces such as weight. Application of the divergence theorem and considering only a portion of the body yields the local form in spatial and material frame

$$\operatorname{div} \boldsymbol{\sigma} + \rho \mathbf{b}_0 = \rho \ddot{\mathbf{x}} \quad \text{and} \quad \operatorname{Div} \boldsymbol{P} + \rho_0 \mathbf{b}_0 = \rho_0 \ddot{\mathbf{x}} \qquad (2.38)$$

which is also known as *Cauchy's First Equation of Motion*.

Balance of angular momentum similarly postulates that the sum of all externally acting moments is equal to the material time derivative of the angular momentum associated to a fixed point of origin:

$$\frac{d}{dt} \int_{\mathcal{B}_t} \rho \mathbf{x} \times \dot{\mathbf{x}} \, dv = \int_{\mathcal{B}_t} \rho \mathbf{x} \times \mathbf{b}_0 \, dv + \int_{\partial \mathcal{B}_t} \mathbf{x} \times \mathbf{t} \, da. \qquad (2.39)$$

We obtain the local form by applying the divergence theorem and considering only a local portion

$$\boldsymbol{\sigma}^{\mathrm{T}} = \boldsymbol{\sigma} \qquad (2.40)$$

which is known as *Cauchy's Second Equation of Motion*. We note that additionally to the Cauchy stress $\boldsymbol{\sigma}$ the second Piola-Kirchhoff stress $\boldsymbol{S}$ and the Kirchhoff stress $\boldsymbol{\tau}$ are symmetric, whereas the first Piola-Kirchhoff stress $\boldsymbol{P}$ is not.

Balance of energy in mechanical systems requires that the change in total energy of a body equals the power applied to a body. We consider only mechanical and thermal energies and subdivide into internal and kinetic energy. The balance equation reads

$$\frac{d}{dt} \int_{\mathcal{B}_t} \rho(e + \tfrac{1}{2}\dot{\mathbf{x}} \cdot \dot{\mathbf{x}}) \, dv = \int_{\mathcal{B}_t} \rho(\mathbf{b}_0 \cdot \dot{\mathbf{x}} + r) \, dv + \int_{\partial \mathcal{B}_t} (\mathbf{t} \cdot \dot{\mathbf{x}} - \mathbf{q} \cdot \mathbf{n}) \, da \qquad (2.41)$$

with $e = e(\mathbf{x}, t)$ representing the internal energy per unit reference mass, $r = r(\mathbf{x}, t)$ the external heat source per unit reference mass, and $\mathbf{q}$ the spatial heat flux. Note that $\mathbf{n}$ is pointing towards the outside of the body and thus $-\mathbf{q} \cdot \mathbf{n}$ represents heat supply to the body. The energy balance is equivalent to the *first law of thermodynamics*.

The local spatial form

$$\rho \dot{e} = \boldsymbol{\sigma} : \boldsymbol{d} + \rho r - \operatorname{div} \mathbf{q} \tag{2.42}$$

is obtained via the divergence theorem, consideration of a local portion and with the local balance of linear momentum. Its local material from reads

$$\rho_0 \dot{e} = \boldsymbol{S} : \dot{\boldsymbol{E}} + \rho_0 r - \operatorname{Div} \mathbf{Q} \tag{2.43}$$

where $\mathbf{Q} = J\boldsymbol{F}^{-1} \cdot \mathbf{q}$. One may observe that there are work conjugate pairs of stress and strain rate and each form of internal mechanical power

$$\mathcal{P}_{\text{int}} = \int_{\mathcal{B}_t} \boldsymbol{\sigma} : \boldsymbol{d} \, \mathrm{d}v = \int_{\mathcal{B}_0} \boldsymbol{P} : \dot{\boldsymbol{F}} \, \mathrm{d}V = \int_{\mathcal{B}_0} \boldsymbol{S} : \dot{\boldsymbol{E}} \, \mathrm{d}V \tag{2.44}$$

is equivalent.

Entropy Inequality also known as *second law of thermodynamics* or as *Clausius-Duhem inequality* postulates that the direction of a thermodynamical process is naturally determined. For instance, heat flux is always directed from the warmer to the colder medium. Thus the temporal change in total entropy $\eta(\mathbf{x}, t)$ is always larger or equal than the entropy difference between supply caused by external heat production and supply caused by heat flux over the body surface. The integral form reads

$$\frac{\mathrm{d}}{\mathrm{d}t} \int_{\mathcal{B}_t} \rho \eta \, \mathrm{d}v \geq \int_{\mathcal{B}_t} \frac{\rho r}{\vartheta} \, \mathrm{d}v - \int_{\partial \mathcal{B}_t} \frac{\mathbf{q} \cdot \mathbf{n}}{\vartheta} \tag{2.45}$$

with the absolute temperature $\vartheta = \vartheta(\mathbf{x}, t)$. With application of the divergence theorem and consideration of a local portion and the energy balance one obtains the local form in spatial frame

$$\rho \vartheta \dot{\eta} - \rho \dot{e} + \boldsymbol{\sigma} : \boldsymbol{d} - \frac{1}{\vartheta} \mathbf{q} \cdot \operatorname{grad} \vartheta \geq 0. \tag{2.46}$$

We introduce the *(Helmholtz) free energy* ψ defined as $\psi := e - \vartheta \eta$ and reformulate the entropy inequality

$$\mathcal{D} := -\rho(\dot{\psi} + \eta \dot{\vartheta}) + \boldsymbol{\sigma} : \boldsymbol{d} - \frac{1}{\vartheta} \mathbf{q} \cdot \operatorname{grad} \vartheta \geq 0, \tag{2.47}$$

with $\mathcal{D}$ as dissipation per unit reference mass. If we consider only isothermal processes with constant temperature we obtain

$$\mathcal{D} = \boldsymbol{\sigma} : \boldsymbol{d} - \rho \dot{\psi} \geq 0 \tag{2.48}$$

which may be transferred to material frame yielding

$$\mathcal{D} = \boldsymbol{S} : \dot{\boldsymbol{E}} - \rho_0 \dot{\psi} \geq 0 \tag{2.49}$$

The Helmholtz energy ψ is defined with respect to unit reference mass. The free energy Ψ with respect to unit reference volume $\Psi = \rho_0 \psi$ completely characterizes material properties including the referential density and therefore serves as important quantity wherefrom constitutive models will be derived in Chapter 4.

In the case of a purely elastic process dissipation vanishes and equality holds. With $\dot{\Psi} = \frac{\partial \Psi}{\partial \boldsymbol{E}} : \dot{\boldsymbol{E}}$ Equation (2.49) yields

$$\boldsymbol{S} : \dot{\boldsymbol{E}} = \frac{\partial \Psi}{\partial \boldsymbol{E}} : \dot{\boldsymbol{E}} \tag{2.50}$$

which has to hold for arbitrary processes and we arrive at

$$\boldsymbol{S} = \frac{\partial \Psi}{\partial \boldsymbol{E}} = 2 \frac{\partial \Psi}{\partial \boldsymbol{C}} \tag{2.51}$$

2.4 Initial Boundary Value Problem

The initial boundary value problem of continuum mechanics consists of a set of coupled partial differential equations satisfying specified boundary and initial conditions. Formulated in material frame, the local equilibrium equation in strong form follows from Equation (2.37) to

$$\text{Div}\, \boldsymbol{P} + \rho_0 \mathbf{b}_0 = \rho_0 \ddot{\mathbf{x}} \qquad \text{in } \mathcal{B}_0 \times [t_0, t_E]. \tag{2.52}$$

It is essential to specify *boundary conditions* on the considered body (or domain), separated in *Neumann boundary conditions* (or force boundary conditions)

$$\boldsymbol{P} \cdot \boldsymbol{N} = \hat{\boldsymbol{t}}_0 \qquad \text{on } \Gamma_N = \partial \mathcal{B}_{0;N} \times [t_0, t_E] \tag{2.53}$$

and *Dirichlet boundary conditions* (or displacement boundary conditions)

$$\mathbf{u} = \hat{\mathbf{u}} \qquad \text{on } \Gamma_D = \partial \mathcal{B}_{0;D} \times [t_0, t_E] \tag{2.54}$$

where

$$\Gamma_N \cap \Gamma_D = \emptyset \quad \text{and} \quad \Gamma_N \cup \Gamma_D = \partial \mathcal{B}_0 \tag{2.55}$$

must hold. To completely define the time dependency *initial conditions* have to be specified

$$\mathbf{u}(\mathbf{X}, t_o) = \hat{\mathbf{u}}_0(\mathbf{X}) \qquad \text{in } \mathcal{B}_0 \tag{2.56}$$
$$\dot{\mathbf{x}}(\mathbf{X}, t_o) = \hat{\mathbf{x}}_0(\mathbf{X}) \qquad \text{in } \mathcal{B}_0 \tag{2.57}$$

The definition of a constitutive equation according to (2.51) completes the initial boundary value problem. We will solve these problem types by means of the finite element method as discussed in detail in the following chapter.

Remark In the remainder of the present work the considered problems are sufficiently slow and start from a position of rest, so that inertia forces play a subsidiary role. They are therefore neglected and the quasi-static response is examined. The time t becomes merely an algorithmic parameter describing the evolution of the considered process. This is also beneficial, as issues of numerical time integration do not have to be considered and we refer to the literature for further details, for instance the books of Hughes [147], Crisfield [62], Wriggers [333], and Belytschko *et al.* [28].

3. Efficient Finite Elements for Arterial Wall Modeling

The simulation of patient-specific arterial wall stresses is performed with the finite element method. The involved complex geometries are inevitably three-dimensional and efficiently modeled as thin-walled structures. It is well-known that standard finite elements are characterized by poor performance within the incompressible regime and bending-dominated problems. Therefore, special element technology is usually employed to overcome these issues. Within the present chapter, the necessary prerequisites of nonlinear finite element analysis are shortly reviewed. Specific locking phenomena are described and the limitations of element design are discussed. In addition, the theoretical background of popular methods to tackle locking is reported. Subsequently, we propose a set of advanced finite elements meeting the requirements of arterial wall modeling, including bulky 8-node hexahedrons, 8-node solid-shells, and wedge-shaped 6-node solid-shells. Popular benchmark examples are studied to investigate the performance of the proposed elements.

3.1 Prerequisites of Finite Element Analysis

The initial boundary value problem for nonlinear continuum mechanics, derived in Section 2.4 and in the following restricted to quasi-static problems for convenience, can only in rare cases be solved analytically. Integral or so-called weak formulations and variational principles are the basis for numerical solution techniques such as the finite element method. Approximation of the involved functions by means of finite elements can be solved via numerical solution schemes. Unfortunately, standard linear (in terms of their shape functions) finite element approximations are affected by several locking problems.

The popular variational *principle of virtual work (PvW)* is in such cases usually replaced by more general variational principles to derive advanced finite elements which overcome these shortcomings. Linearization of the underlying equations becomes necessary to apply efficient numerical solution techniques such as the Newton-Raphson method distinguished by its quadratic rate of convergence within regions sufficiently close to the solution. In the following, we review two important variational principles followed by the linearization and discretization with finite elements.

3.1.1 Variational Principles

As mentioned, the most common variational principle to derive finite elements is the principle of virtual work. It allows a rather intuitive physical interpretation, namely that the work done by internal stresses on virtual strains balances the work done by external forces on virtual displacements. Only one field, the displacements, is involved and the so-called displacement-based finite elements are derived therefrom, as presented in the following.

In order to overcome the performance issues the more general three-field *Hu-Washizu principle* is employed to formulate advanced or so-called (hybrid) mixed elements. Note that according to Felippa [82] this principle should rather be termed *Fraeijs de Veubeke-Hu-Washizu principle (VHW)*. Therein, two additional fields, the stresses and the strains, are involved. A modified form of the VHW principle is employed for a popular element technology, the *Enhanced Assumed Strain* (EAS) method, proposed by Simo and Rifai [272]. Another well known principle is the two-field *Hellinger-Reissner principle* where besides the displacement field also the stress field is involved.

Principle of Virtual Work

This principle is based on the physically intuitive interpretation that at a system in equilibrium any kinematically admissible, infinitesimal, virtual displacement $\delta \mathbf{u}$ performs no work on that system. The PvW in material frame results from multiplying the local equilibrium equation (2.52) together with the force boundary conditions (FBC) with virtual displacements $\delta \mathbf{u}$ which satisfy the Dirichlet boundary conditions (DBC). They can be also interpreted as vector valued test function or weighting function within the method of weighted residuals. Integration over the corresponding domains yields

$$\int_{\mathcal{B}_0} [\mathrm{Div}(\underbrace{\boldsymbol{FS}}_{\boldsymbol{P}}) + \rho_0 \mathbf{b}_0] \cdot \delta \mathbf{u} \; \mathrm{d}V + \int_{\Gamma_N} [\hat{\mathbf{t}}_0 - \boldsymbol{FS} \cdot \mathbf{N}] \cdot \delta \mathbf{u} \; \mathrm{d}A = 0. \tag{3.1}$$

We omit the index $(\,\cdot\,)_0$ designating the material frame from now on for the sake of compactness. By making use of the Gauss divergence theorem, the symmetry of $\boldsymbol{S}$ and the variation of the Green-Lagrange strains

$$\delta \boldsymbol{E} = \frac{1}{2} \left((\boldsymbol{F}^{\mathrm{T}} \,\mathrm{Grad}\, \delta \mathbf{u})^{\mathrm{T}} + \boldsymbol{F}^{\mathrm{T}} \,\mathrm{Grad}\, \delta \mathbf{u} \right), \tag{3.2}$$

we obtain the material description of the PvW, that is, the virtual work of internal and external forces have to balance each other, their difference has to vanish

$$\delta \Pi_{\mathrm{PvW}}(\mathbf{u}, \delta \mathbf{u}) = \underbrace{\int_{\mathcal{B}_0} \boldsymbol{S} : \delta \boldsymbol{E} \; \mathrm{d}V}_{\delta \Pi_{\mathrm{PvW}}^{\mathrm{int}}} \underbrace{- \int_{\mathcal{B}_0} \varrho \mathbf{b} \cdot \delta \mathbf{u} \; \mathrm{d}V - \int_{\Gamma_N} \left(\hat{\mathbf{t}} - \mathbf{t}^{\mathbf{u}} \right) \cdot \delta \mathbf{u} \; \mathrm{d}A}_{\delta \Pi_{\mathrm{PvW}}^{\mathrm{ext}}} = 0. \tag{3.3}$$

The superscript $(\,\cdot\,)^{\mathbf{u}}$ designates here and from now on a dependency of the displacement $\mathbf{u}$. Only the displacement field $\mathbf{u}$ appears as primary field and the equilibrium and traction boundary conditions are fulfilled in a weak sense. A clarifying illustration as so-called Tonti-diagram, after the Italian mathematician Enzo Tonti [305], is depicted for the PvW in material frame in Fig. 3.1. For a purely elastic problem the same formulation can also be obtained by the principle of stationary potential energy. However, the PvW does not necessitate the existence of a potential and is therefore applicable also for more general problems for instance involving inelastic materials.

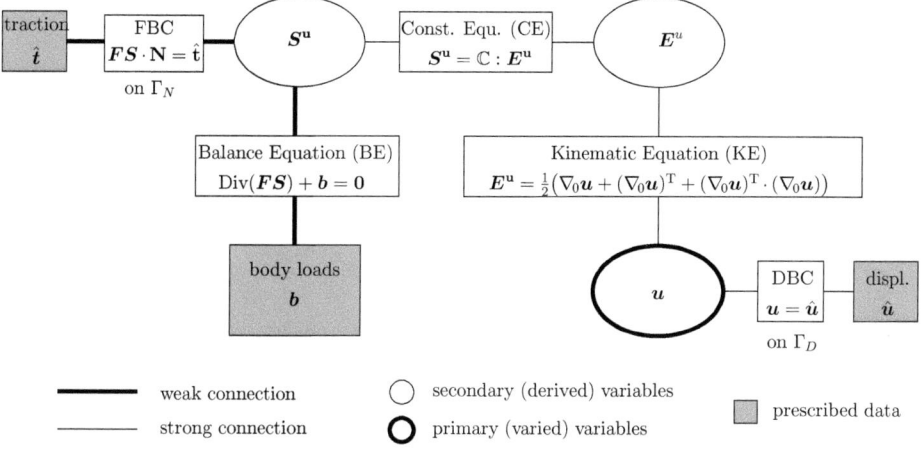

Fig. 3.1: Tonti-diagram of the Principle of Virtual Work in material frame.

Fraeijs de Veubeke-Hu-Washizu principle

Additionally, we introduce the VHW variational principle which serves as point of departure for several advanced finite element methods

$$\Pi_{\text{VHW}}(\mathbf{u}, \boldsymbol{E}, \boldsymbol{S}) = \int_{\mathcal{B}_0} [\Psi(\boldsymbol{E}) + \boldsymbol{S} : (\boldsymbol{E}^{\mathbf{u}} - \boldsymbol{E}) - \varrho \mathbf{b} \cdot \mathbf{u}] \, \text{d}V$$

$$- \int_{\Gamma_N} \hat{\mathbf{t}} \cdot \mathbf{u} \, \text{d}A + \int_{\Gamma_D} \mathbf{t}^S \cdot (\hat{\mathbf{u}} - \mathbf{u}) \, \text{d}A \quad \rightarrow \quad \text{stat.} \quad (3.4)$$

Invoking stationary condition, Gauss divergence theorem, and symmetry of $\boldsymbol{S}$ yields

$$\delta\Pi_{\text{VHW}}(\mathbf{u}, \boldsymbol{E}, \boldsymbol{S}) = \int_{\mathcal{B}_0} \left[\overbrace{\frac{\partial \Psi}{\partial \boldsymbol{E}}}^{\boldsymbol{S}^E} : \delta \boldsymbol{E} + \delta \boldsymbol{S} : (\boldsymbol{E}^{\mathbf{u}} - \boldsymbol{E}) \right] \text{d}V$$

$$- \int_{\mathcal{B}_0} [\delta \mathbf{u} \cdot \text{Div}(\boldsymbol{FS}) + \boldsymbol{S} : \delta \boldsymbol{E} + \varrho \mathbf{b} \cdot \delta \mathbf{u}] \, \text{d}V$$

$$- \int_{\Gamma_N} (\hat{\mathbf{t}} - \mathbf{t}^S) \cdot \delta \mathbf{u} \, \text{d}A + \int_{\Gamma_D} \delta \mathbf{t}^S \cdot (\hat{\mathbf{u}} - \mathbf{u}) \, \text{d}A = 0. \quad (3.5)$$

Herein, the displacement field **u**, the stress field **S** and the strain field **E** appear as primary variables. All related field equations as well as the boundary conditions are satisfied only in a weak sense. Note, that we have introduced in Equation (3.4) the strain energy function $\Psi(\boldsymbol{E})$. Nonetheless, the principle is also valid for inelastic problems in which the potential for the constitutive relation is undefined. In this case we remain with a weak satisfaction of the weighted residual equation (SS): $\boldsymbol{S}^E - \boldsymbol{S}$. Equivalently, the primary variable $\boldsymbol{E}$ yields another weighted residual (EE): $\boldsymbol{E}^u - \boldsymbol{E}$. Fig. 3.2 depicts the Tonti-diagram for the VHW variational principle. A different point of view suggests interpreting the primary stress and strain fields as Lagrangean multiplier fields (see Bischoff [34] for details).

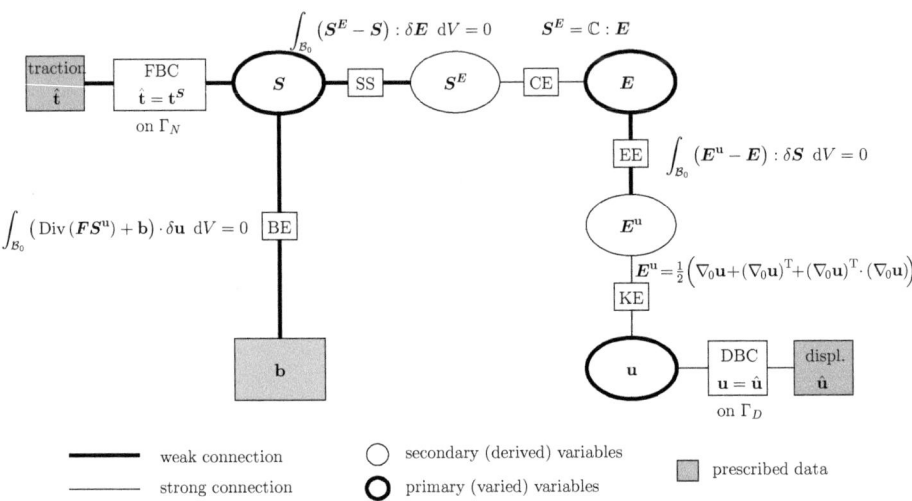

Fig. 3.2: Tonti-diagram of the VHW variational principle in material frame.

3.1.2 Linearization of the Virtual Work Principle

Several sources of nonlinearity suggest a linearization of the underlying equations to efficiently solve the problem using a Newton-Raphson procedure with the beneficial property of a quadratic rate of convergence in a sufficiently small neighborhood of the solution. Though it is also possible to discretize the nonlinear equations and then linearize with respect to the nodal unknowns, we rather linearize the previously introduced variational formulations and then discretize the resulting equations. The linearization of the PvW (Equation (3.3)) at $\mathbf{u} = \bar{\mathbf{u}}$ is defined as

$$\text{Lin}\,\delta\Pi_{\text{PvW}}(\bar{\mathbf{u}}, \delta\mathbf{u}, \Delta\mathbf{u}) := \delta\Pi_{\text{PvW}}(\bar{\mathbf{u}}, \delta\mathbf{u}) + \Delta\delta\Pi_{\text{PvW}}(\bar{\mathbf{u}}, \delta\mathbf{u}, \Delta\mathbf{u}). \tag{3.6}$$

The incremental virtual work $\Delta\delta\Pi_{\text{PvW}}$ is obtained through the directional derivative (Gâteaux derivative), given as

$$\Delta\delta\Pi_{\text{PvW}}(\bar{\mathbf{u}}, \delta\mathbf{u}, \Delta\mathbf{u}) = \frac{\text{d}}{\text{d}\epsilon}[\delta\Pi_{\text{PvW}}(\bar{\mathbf{u}} + \epsilon\Delta\mathbf{u}, \delta\mathbf{u})]\bigg|_{\epsilon=0} = \text{D}\,\delta\Pi_{\text{PvW}}(\bar{\mathbf{u}}, \delta\mathbf{u})\cdot\Delta\mathbf{u} \quad (3.7)$$

and we may split into internal and external contribution yielding

$$\text{D}\,\delta\Pi_{\text{PvW}}(\bar{\mathbf{u}}, \delta\mathbf{u})\cdot\Delta\mathbf{u} = \text{D}\,\delta\Pi^{\text{int}}_{\text{PvW}}(\bar{\mathbf{u}}, \delta\mathbf{u})\cdot\Delta\mathbf{u} - \text{D}\,\delta\Pi^{\text{ext}}_{\text{PvW}}(\bar{\mathbf{u}}, \delta\mathbf{u})\cdot\Delta\mathbf{u}. \quad (3.8)$$

The external virtual work may have contributions from body forces **b** and surface tractions **t**. Body forces rarely depend on the deformation, but there exist a wide variety of different traction forces and some do depend on the deformation. A classical example is a closed structure under internal pressure leading to a surface load acting always orthogonally to the surface. To achieve a fully quadratic rate of convergence the corresponding contribution to the external virtual work needs to be consistently linearized. We skip the derivation and restrict ourselves here to displacement-independent forces. Details may be found in the literature, for instance Bonet and Wood [45]. Thus, the directional derivative of the external virtual work vanishes: $\text{D}\,\tilde{\Pi}^{\text{ext}}_{\text{PvW}} = 0$.

The internal virtual work $\delta\Pi^{\text{int}}_{\text{PvW}} = \int_{\mathcal{B}_0} \mathbf{S}:\delta\mathbf{E}\,\text{d}V$ is linearized using the product rule for the directional derivative, yielding

$$\begin{aligned}
\text{D}\,\delta\Pi^{\text{int}}_{\text{PvW}}(\bar{\mathbf{u}}, \delta\mathbf{u})\cdot\Delta\mathbf{u} &= \int_{\mathcal{B}_0} \text{D}(\mathbf{S}:\delta\mathbf{E})\,\text{d}V \\
&= \int_{\mathcal{B}_0} \delta\mathbf{E}:\underbrace{\text{D}\,\mathbf{S}\cdot\Delta\mathbf{u}}_{\Delta\mathbf{S}}\,\text{d}V + \int_{\mathcal{B}_0} \mathbf{S}:\text{D}\,\delta\mathbf{E}\cdot\Delta\mathbf{u}\,\text{d}V \\
&= \int_{\mathcal{B}_0} \delta\mathbf{E}:\mathbb{C}:\underbrace{\text{D}\,\mathbf{E}\cdot\Delta\mathbf{u}}_{\Delta\mathbf{E}}\,\text{d}V + \int_{\mathcal{B}_0} \mathbf{S}:\underbrace{\text{D}\,\delta\mathbf{E}\cdot\Delta\mathbf{u}}_{\Delta\delta\mathbf{E}}\,\text{d}V,
\end{aligned} \quad (3.9)$$

where we inserted the linearization of the second Piola-Kirchhoff stresses

$$\Delta\mathbf{S} = \mathbb{C}:\text{D}\,\mathbf{E}\cdot\Delta\mathbf{u} \quad (3.10)$$

with the definition of the material elasticity tensor $\mathbb{C} = 2\frac{\partial\mathbf{S}}{\partial\mathbf{C}}$. The virtual Green-Lagrange strain tensor, its increment and the linearized Green-Lagrange strain are given as

$$\delta\mathbf{E} = \tfrac{1}{2}((\nabla_0\delta\mathbf{u})^{\text{T}}\mathbf{F} + \mathbf{F}^{\text{T}}(\nabla_0\delta\mathbf{u})) \quad (3.11)$$

$$\Delta\delta\mathbf{E} = \tfrac{1}{2}\left((\nabla_0\delta\mathbf{u})^{\text{T}}\nabla_0(\Delta\mathbf{u}) + (\nabla_0(\Delta\mathbf{u}))^{\text{T}}\nabla_0\delta\mathbf{u}\right) \quad (3.12)$$

$$\Delta\mathbf{E} = \tfrac{1}{2}((\nabla_0(\Delta\mathbf{u}))^{\text{T}}\mathbf{F} + \mathbf{F}^{\text{T}}(\nabla_0(\Delta\mathbf{u}))), \quad (3.13)$$

with $\nabla_0\delta\mathbf{u} = \text{Grad}\,\delta\mathbf{u}$ and $\nabla_0(\Delta\mathbf{u}) = \text{Grad}\,\Delta\mathbf{u}$. Note also that the variation of the current configuration equals the variation of the displacement $\delta(\mathbf{x} = \mathbf{X} + \mathbf{u}) = \delta\mathbf{u}$.

It is remarked that in case of linear kinematics the well-known linear strains are recovered, since the deformation gradient $\mathbf{F}$ in the *undeformed* configuration is the identity tensor and the incremental displacements $\Delta\mathbf{u}$ are identical to the ordinary displacements $\mathbf{u}$, leading to

$$\text{D}\,\mathbf{E}(\mathbf{u}_0)\cdot\mathbf{u} = \tfrac{1}{2}(\nabla\mathbf{u} + (\nabla\mathbf{u})^{\text{T}}) \quad (3.14)$$

For the linearization of the VHW principle we refer to Section 3.4.3.

3.1.3 Discretization of the Virtual Work Principle

The numerical solution by means of the finite element method requires the discretization of the equations presented so far. Thereby, the continuous functions are approximated with assumed ansatz- or shape-functions, typically low-order polynomials, and associated discrete nodes. The problem is transferred from finding unknown functions $\mathbf{u}$ and $\delta\mathbf{u}$ to finding discrete unknowns composed in the vectors $\mathbf{D}, \delta\mathbf{D}$. A key of the finite element concept is to combine a number of nodes to a sub domain (element) and define the shape functions only locally within one element (local support). Hence, the involved integrals need to be evaluated only within each element and the full domain $\mathcal{B}_0$ is by the union of all element domains $\mathcal{B}_{(e)}$ which are not allowed to overlap

$$\mathcal{B}_0 \approx \mathcal{B}_0^h = \bigcup_{e=1}^{n_{ele}} \mathcal{B}_{(e)} \tag{3.15}$$

The unknown field $\mathbf{u}$ is approximated by a set of suitable shape functions defined within one element, composed in the vector $\mathbf{N}$, and the corresponding discrete nodal unknowns $\mathbf{d}$:

$$\mathbf{u} \approx \mathbf{u}^h = \bigcup_{e=1}^{n_{ele}} \mathbf{u}_{(e)}^h \quad \text{with} \quad \mathbf{u}_{(e)}^h = \mathbf{N}\,\mathbf{d} \tag{3.16}$$

The so-called Bubnov-Galerkin approach suggests that the same shape functions are used for the approximation of the virtual displacements $\delta\mathbf{u}_{(e)}^h = \mathbf{N}\,\delta\mathbf{d}$ as well as for the incremental displacements $\Delta\mathbf{u}_{(e)}^h = \mathbf{N}\,\Delta\mathbf{d}$. Introducing the local element coordinate system (parameter space) (ξ, η, ζ) with $\xi, \eta, \zeta \in [-1; 1]$ we may write

$$\mathbf{u}_{(e)}(\xi, \eta, \zeta) = \sum_{I=1}^{n_{nd}} N_I(\xi, \eta, \zeta)\,\mathbf{d}_I, \tag{3.17}$$

where n_{nd} is the number of nodes per element and I is the index of one particular node. For convenience we restrict ourselves from now on to the most popular linear shape functions. Then the nodal unknowns can be directly identified with nodal displacements at the discrete locations $\mathbf{X}_I$.

Following the isoparametric concept the domain is approximated using again the same shape functions. This yields the following for every point $\mathbf{X}$ in reference and current configuration:

$$\mathbf{X} \approx \mathbf{X}^h = \bigcup_{e=1}^{n_{ele}} \mathbf{X}_{(e)}^h \quad \text{and} \quad \mathbf{X}_{(e)}^h = \sum_{I=1}^{n_{nd}} N_I(\xi, \eta, \zeta)\,\bar{\mathbf{X}}_I = \mathbf{N}\,\bar{\mathbf{X}} \tag{3.18}$$

$$\mathbf{x} \approx \mathbf{x}^h = \bigcup_{e=1}^{n_{ele}} \mathbf{x}_{(e)}^h \quad \text{and} \quad \mathbf{x}_{(e)}^h = \sum_{I=1}^{n_{nd}} N_I(\xi, \eta, \zeta)\,\bar{\mathbf{x}}_I = \mathbf{N}\,\bar{\mathbf{x}} \tag{3.19}$$

where $\bar{\mathbf{X}}$ and $\bar{\mathbf{x}}$ represent the vectors of the nodal coordinates of each element in reference and current configuration, respectively. Hence, the mapping $\mathbf{X}(\boldsymbol{\xi})$ between the real space and the parameter space is defined with the vector composition $\boldsymbol{\xi} = [\xi, \eta, \zeta]^T$. The corresponding Jacobian matrix and its inverse are given as

$$\mathbf{J} = \frac{\partial \mathbf{X}}{\partial \boldsymbol{\xi}} \quad \text{and} \quad \mathbf{J}^{-1} = \frac{\partial \boldsymbol{\xi}}{\partial \mathbf{X}} \tag{3.20}$$

3. Efficient Finite Elements for Arterial Wall Modeling

and thus the discretized quantities on the element level are evaluated, as for example

$$\text{Grad } \mathbf{u}_{(e)} = \sum_{I=1}^{n_{nd}} \text{Grad}[N_I] \, \mathbf{d}_I = \sum_{I=1}^{n_{nd}} \frac{\partial N_I}{\partial \boldsymbol{\xi}} \frac{\partial \boldsymbol{\xi}}{\partial \mathbf{X}} \, \mathbf{d}_I = \sum_{I=1}^{n_{nd}} \mathbf{J}^{-T} \frac{\partial N_I}{\partial \boldsymbol{\xi}} \, \mathbf{d}_I. \qquad (3.21)$$

The Jacobian mapping plays an essential factor in the approximation and convergence quality of finite elements. Mind that it becomes nonlinear for distorted elements in material frame and in this case elemental quantities cannot be integrated exactly by common Gauss quadrature. In Fig. 3.3 the Jacobian and the deformation mapping are illustrated for a two-dimensional element.

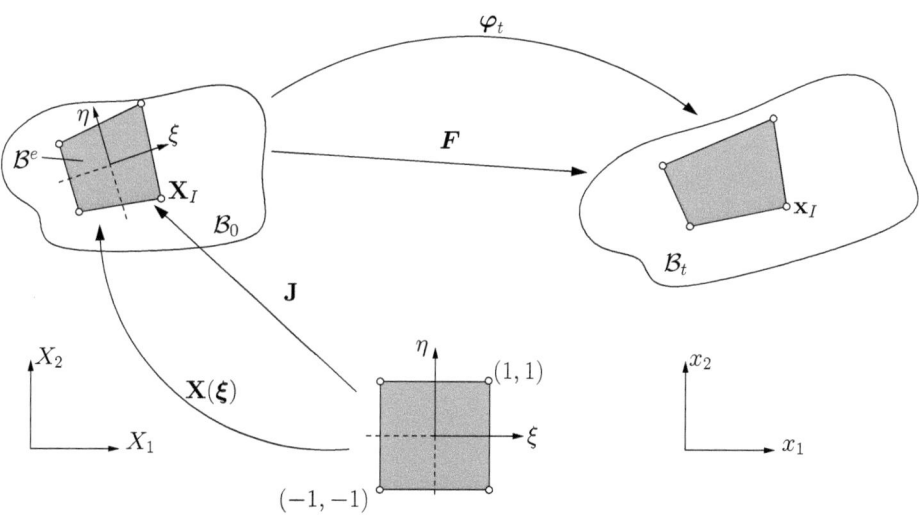

Fig. 3.3: Sketch of deformation and parameter space mapping of a four-noded element.

The next step in deriving the finite element discretization is the introduction of the discrete

Green-Lagrange strain in the so-called *Voigt* vector notation

$$\delta \mathbf{E}_{(e)} = \begin{bmatrix} \delta E_{11} \\ \delta E_{22} \\ \delta E_{33} \\ 2\delta E_{12} \\ 2\delta E_{23} \\ 2\delta E_{13} \end{bmatrix} = \sum_{I=1}^{n_{nd}} \mathbf{B}_I \, \delta \mathbf{d}_I \quad \text{with}$$

$$\mathbf{B}_I = \begin{bmatrix} F_{11}N_{I,1} & F_{21}N_{I,1} & F_{31}N_{I,1} \\ F_{12}N_{I,2} & F_{22}N_{I,2} & F_{32}N_{I,2} \\ F_{13}N_{I,3} & F_{23}N_{I,3} & F_{33}N_{I,3} \\ F_{11}N_{I,2} + F_{12}N_{I,1} & F_{21}N_{I,2} + F_{22}N_{I,1} & F_{31}N_{I,2} + F_{32}N_{I,1} \\ F_{12}N_{I,3} + F_{13}N_{I,2} & F_{22}N_{I,3} + F_{23}N_{I,2} & F_{32}N_{I,3} + F_{33}N_{I,2} \\ F_{11}N_{I,3} + F_{13}N_{I,1} & F_{21}N_{I,3} + F_{23}N_{I,1} & F_{31}N_{I,3} + F_{33}N_{I,1} \end{bmatrix} \quad \text{and}$$

$$\delta \mathbf{d}_I = \begin{bmatrix} \delta u_{1I} \\ \delta u_{2I} \\ \delta u_{3I} \end{bmatrix}. \tag{3.22}$$

The nodal components can be arranged as vector and matrix, respectively, such that

$$\mathbf{B} = [\mathbf{B}_1 \mid \mathbf{B}_2 \mid \mathbf{B}_3 \mid \cdots \mid \mathbf{B}_{n_{nd}}] \tag{3.23}$$

$$\delta \mathbf{d} = [\mathbf{d}_1^{\mathrm{T}} \mid \mathbf{d}_2^{\mathrm{T}} \mid \mathbf{d}_3^{\mathrm{T}} \mid \cdots \mid \mathbf{d}_{n_{nd}}^{\mathrm{T}}]^{\mathrm{T}}. \tag{3.24}$$

Moreover, the second Piola-Kirchhoff stress on element level is defined in Voigt notation as the vector

$$\mathbf{S}_{(e)} = \begin{bmatrix} S_{11} & S_{22} & S_{33} & S_{12} & S_{23} & S_{13} \end{bmatrix}^{\mathrm{T}} \tag{3.25}$$

Then, at the element level the discretized internal virtual work is obtained as

$$\int_{\mathcal{B}_0^{(e)}} \delta \mathbf{E}_{(e)}^{\mathrm{T}} \, \mathbf{S}_{(e)} \, \mathrm{d}V = \int_{\mathcal{B}_0^{(e)}} (\mathbf{B} \, \delta \mathbf{d})^{\mathrm{T}} \, \mathbf{S}_{(e)} \, \mathrm{d}V = \delta \mathbf{d}^{\mathrm{T}} \int_{\mathcal{B}_0^{(e)}} \mathbf{B}^{\mathrm{T}} \, \mathbf{S}_{(e)} \, \mathrm{d}V. \tag{3.26}$$

The integrals over the element domain $\mathcal{B}_{(e)}$ have in general no analytic solution and numerical integration schemes are employed. We apply the standard Gauss integration rule which replaces the integrals by a sum of integrand functions evaluated at specified Gauss points multiplied with certain weighting factors. We refer to the literature for details, for instance the books by Hughes [147] or Zienkiewicz *et al.* [342].

Integration over the full domain $\mathcal{B}_0$ is finally obtained via the *assembly* operator governing the arrangement of elemental quantities into the global equation system:

$$\int_{\mathcal{B}_0} (\,\cdot\,) \, \mathrm{d}V \approx \mathop{\mathbf{A}}_{e=1}^{n_{ele}} \int_{\mathcal{B}_0^{(e)}} (\,\cdot\,) \, \mathrm{d}V. \tag{3.27}$$

The discrete version of the internal virtual work reads finally:

$$\delta\Pi_{\text{PvW}}^{\text{int}} = \underset{e=1}{\overset{n_{ele}}{\mathbf{A}}} \delta\mathbf{d}^{\text{T}} \int_{\mathcal{B}_0^{(e)}} \mathbf{B}^{\text{T}} \mathbf{S}_{(e)} \, dV$$

$$= \delta\mathbf{D}^{\text{T}} \underbrace{\underset{e=1}{\overset{n_{ele}}{\mathbf{A}}} \int_{\mathcal{B}_0^{(e)}} \mathbf{B}^{\text{T}} \mathbf{S}_{(e)} \, dV}_{\mathbf{f}_{\text{int}}} := \mathbf{F}_{\text{int}}(\mathbf{u}). \quad (3.28)$$

Together with the discretized form of the external virtual work $\delta\Pi_{\text{PvW}}^{\text{ext}} := \mathbf{F}_{\text{ext}}$ we arrive at the system of nonlinear equations

$$\mathbf{R} = \mathbf{F}_{\text{int}}(\mathbf{D}) - \mathbf{F}_{\text{ext}} = \mathbf{0} \quad (3.29)$$

To solve this efficiently by means of an iterative Newton-Raphson scheme we need the discrete linearized form of equation (3.29) which is at the same time the discretization of the internal virtual work, Equation (3.9). Linearization around a specific displacement state $\mathbf{D}^i$ at iteration step i reads

$$\text{Lin } \mathbf{R} = \underbrace{\mathbf{R}(\mathbf{D}^i)}_{\text{RHS}} + \underbrace{\left.\frac{\partial \mathbf{R}(\mathbf{D})}{\partial \mathbf{D}}\right|_{\mathbf{D}^i}}_{\text{tangent } \mathbf{K}} \Delta\mathbf{D}^{i+1} = \mathbf{0}. \quad (3.30)$$

The tangent (stiffness) matrix $\mathbf{K}$ is the assembly of the element matrices $\mathbf{k}_{(e)}$

$$\mathbf{K} = \underset{e=1}{\overset{n_{ele}}{\mathbf{A}}} \mathbf{k}_{(e)} \quad (3.31)$$

and one element matrix $\mathbf{k}_{(e)}$ is obtained from a discretization of the linearized internal virtual work at element level $\left(D\, \delta\Pi_{\text{PvW}}^{\text{int}} \cdot \Delta\mathbf{u}\right)_{(e)}$ given as

$$\int_{\mathcal{B}_0^{(e)}} D\,\mathbf{E}\cdot\delta\mathbf{u} : \mathbb{C} : D\,\mathbf{E}\cdot\Delta\mathbf{u} \, dV + \int_{\mathcal{B}_0^{(e)}} \mathbf{S} : \left((\nabla_0(\Delta\mathbf{u}))^{\text{T}}\nabla_0\delta\mathbf{u}\right) \, dV =$$

$$\delta\mathbf{d}^{\text{T}} \underbrace{\int_{\mathcal{B}_0^{(e)}} \mathbf{B}^{\text{T}}\bar{\mathbb{C}}\,\mathbf{B} \, dV}_{\mathbf{k}_{\text{eu}}} \Delta\mathbf{d} + \delta\mathbf{d}^{\text{T}} \mathbf{k}_{\text{geo}} \Delta\mathbf{d}. \quad (3.32)$$

The element stiffness matrix $\mathbf{k}_{(e)}$ is herein split into two parts, the elastic and initial displacement contribution $\mathbf{k}_{\text{eu}}$ and the geometric contribution $\mathbf{k}_{\text{geo}}$. For a convenient and efficient matrix notation of $\mathbf{k}_{\text{eu}}$ we introduce the 6-by-6 matrix $\bar{\mathbb{C}}$ as reduced notation of the fourth-order tensor $\mathbb{C}$. A matrix notation for $\mathbf{k}_{\text{geo}}$ is a little cumbersome and we therefore use the index notation, yielding

$$\delta\mathbf{d}^{\text{T}} \mathbf{k}_{\text{geo}} \Delta\mathbf{d} = \sum_{I=1}^{n_{nd}} \sum_{J=1}^{n_{nd}} \delta\mathbf{d}_I^{\text{T}} \mathbf{G}_{IJ} \Delta\mathbf{d}_J \quad (3.33)$$

with $\mathbf{G} = \text{diag}(\hat{S}_{IK}, \hat{S}_{IK}, \hat{S}_{IJ})$ and

$$\begin{aligned}\hat{S}_{IK} =\ & S_{11}N_{I,1}N_{J,1} + S_{22}N_{I,2}N_{J,2} + S_{33}N_{I,3}N_{J,3} \\ & + S_{12}(N_{I,1}N_{J,2} + N_{I,2}N_{J,1}) \\ & + S_{23}(N_{I,2}N_{J,3} + N_{I,3}N_{J,2}) \\ & + S_{13}(N_{I,1}N_{J,3} + N_{I,3}N_{J,1})\end{aligned} \qquad (3.34)$$

Finally, within a Newton-Raphson solution procedure the iterative displacement increment

$$\Delta \mathbf{D}^{i+1} = -\mathbf{K}^{-1}\,\mathbf{R}(\mathbf{D}^i) \qquad (3.35)$$

is evaluated at each iteration step i and the displacements are updated

$$\mathbf{D}^{i+1} :\Leftarrow \mathbf{D}^i + \Delta\mathbf{D}^{i+1} \qquad (3.36)$$

until the residuum (RHS) $\mathbf{R}$ is below a specified tolerance and the solution is converged.

3.2 Locking Phenomena of 3D Elements

In the early days of the finite element method it was already recognized that finite elements based upon the PvW may lead to inaccurate results and exhibit slow convergence. The predictions for displacements and stresses become completely useless in certain extreme cases. This phenomenon, known as *locking*, is characterized by a severe underestimation of the displacements, i.e. the structural response is too stiff. Since the late seventies the term locking has been used (see Hughes *et al.* [150]), reflecting the perception that the elements lock themselves against deformation.

A unique, rigorous definition of locking is nonexistent. It can be stated that

> Locking means the effect of a reduced rate of convergence in dependence of a "critical" parameter. In the infinite limit of this parameter the rate of convergence may be zero.

From a mathematical perspective the reason for locking is ill-conditioning, typical for stiff differential equations. A more mechanically motivated explanation would be the inability of a finite element formulation to represent certain deformation modes without unwanted, "parasitic" strains and/or stresses. There is also a numerical point of view in which Hughes [147] draws up the balance between the number of displacement equations and the number of deformation constraints for the discrete problem.

In the following we summarize different locking effects for three-dimensional structures where we restrict ourselves to linear finite elements in terms of their shape functions. For higher order elements following the approach of the so-called *p*-version finite element method different considerations are relevant, see Düster [76] and references therein.

The dependence of a critical parameter allows the classification of two types of locking phenomena. The first type is the over-stiffening with respect to incompressibility, therefore a material locking type called *volumetric locking* with the bulk modulus κ as driving parameter. For the application in biomechanical problems this becomes very important, as the considered materials are usually incompressible.

The second type are so-called *geometric locking* phenomena where the reason for overstiffening lies in violation of a certain kinematic constraint. The driving parameter is always a geometric measure of the considered structure or rather the underlying element geometry, such as the aspect ratio for *shear locking*. In a strict three-dimensional discretization with standard elements this issue would lead to extremely fine meshes for slender structures which are not acceptable from a computational cost point of view. Therefore, already in the beginning of the finite element method there was a high interest in the development of *structural elements* such as beam-, plate-, or shell-elements to simulate the corresponding widely used engineering structures. Here, more specific locking phenomena can be identified, like for example *transverse shear locking*, or *Poisson-thickness locking*, which then depend also on the related structural parameter. But it is important to remark that the locking phenomena are not a result of the involved dimensional reduction or special beam-, plate-, or shell-models. Rather, they are inherent in the physical problem of analyzing for instance thin-walled structures. For a comprehensive review of modeling thin-walled structures see Bischoff *et al.* [37]. Hence, when using three-dimensional finite elements and reasonable meshes for modeling thin-walled structures, all specific geometric locking effects are present. They have to be tackled with certain element techniques leading to the so-called *solid-shell* concepts described in Sections 3.5.2 and 3.5.3. There is a large amount of literature on locking phenomena and methods to overcome these problems for structural elements. We refer the interested reader for more details and references, especially with respect to shells, to Harnau [123], Koschnick [175], Klinkel [169], Bischoff [34], and Andelfinger [5].

In the following, we are concentrating only on three-dimensional solid elements which, however, might become thin. Relevant locking phenomena are volumetric locking, shear locking, trapezoidal locking, and membrane locking which are explained in the subsequent sections.

3.2.1 Volumetric Locking

The driving parameter for volumetric locking is an increasing bulk modulus ($\kappa \to \infty$) of the material involved. It is associated with Poisson's ratio approaching incompressibility ($\nu \to 0.5$). In this case a valid deformation has to be purely isochoric, satisfying the constraint

$$\mathrm{div}(\mathbf{u}) = 0. \tag{3.37}$$

The issue for finite elements in this regard can be illustrated with a simple two-dimensional small deformation example. We consider an incompressible body in pure bending deformation in the $\xi\eta$-plane under plane-strain conditions, see Fig. 3.4 on the left. The deformation may be described by the displacement state

$$u_X \sim \xi\eta \quad \text{and} \quad u_Y \sim -\tfrac{1}{2}(\xi)^2 - \frac{\nu}{2(1-\nu)}(\eta)^2. \tag{3.38}$$

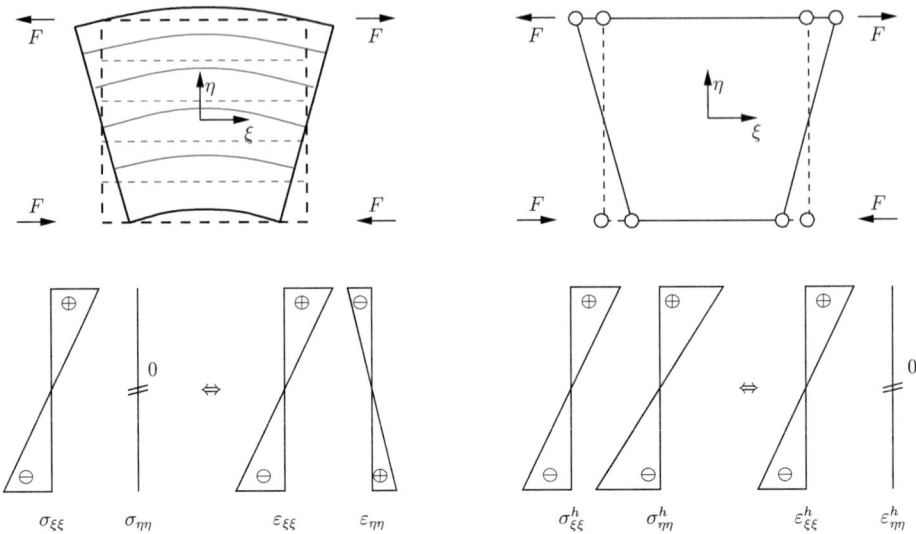

Fig. 3.4: Illustration of volumetric locking.

The corresponding stress distribution of $\sigma_{\xi\xi}$ and $\sigma_{\eta\eta}$ along η is also given. For simplicity we take a St. Venant-Kirchhoff material, implement the plain-strain condition and evaluate the related strain-components

$$\begin{aligned}
\varepsilon_{\xi\xi} &= \eta & \sigma_{\xi\xi} &= \frac{E}{1-\nu^2}\eta \\
\varepsilon_{\eta\eta} &= -\frac{\nu}{1-\nu}\eta & \sigma_{\eta\eta} &= 0 \\
\varepsilon_{\xi\eta} &= 0 & \sigma_{\xi\eta} &= 0.
\end{aligned} \qquad (3.39)$$

Their distribution along η is also plotted below in Fig. 3.4. It is observed that due to the lateral extension the body gets thinner at the top but thicker at the bottom, keeping the overall volume constant but shifting the material center upwards.

Let us now consider the corresponding situation for a finite element, as seen in Fig. 3.4 on the right. The deformation approximation of the single element is not capable of neither reproducing the curved shape nor shifting the material center and the strain state $\varepsilon^h_{\eta\eta}$ is constant zero. With implemented material law this yields

$$\begin{aligned}
\varepsilon^h_{\xi\xi} &= \eta & \sigma^h_{\xi\xi} &= \frac{E\nu(1-\nu)}{(1+\nu)(1-2\nu)}\eta \\
\varepsilon^h_{\eta\eta} &= 0 & \sigma^h_{\eta\eta} &= \frac{E\nu}{(1+\nu)(1-2\nu)}\eta \\
\varepsilon^h_{\xi\eta} &= \xi & \sigma^h_{\xi\eta} &= \frac{E}{2(1+\nu)}\xi.
\end{aligned} \qquad (3.40)$$

In the limit ($\nu \to \frac{1}{2}$) the shear stress tends to a finite value, but the "parasitic" stress $\sigma_{\eta\eta}$ rises to infinity.

The deficit here is that the element is not capable of reproducing the linear stress ε_{22} in the vertical direction. This problem can be extended straightforwardly to tree dimensions. A successful approach to avoid this locking phenomenon is to allow such a strain state by manipulating the element. These ideas were followed in the *incompatible modes* approach by Wilson [331] and the EAS method by Simo and Rifai [272].

3.2.2 Shear Locking

The problem of shear locking is in contrast to the previous locking phenomenon dependent on the element geometry, more precisely on the aspect ratio, and thus a *geometric locking* type. It results from the inability of finite elements to model pure bending deformation without any shear, because both deformations and thus also the stresses are directly coupled through the underlying ansatz, see the illustration in Fig. 3.5.

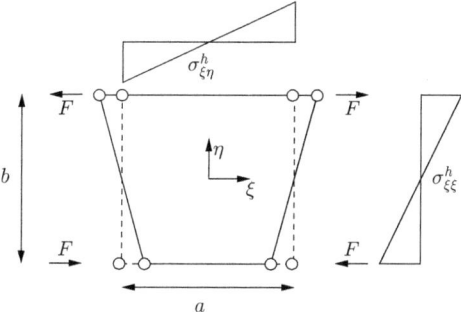

Fig. 3.5: Illustration of shear locking.

The corresponding strain energy related to these "parasitic" shear strains and stresses can be evaluated, exemplified by a two-dimensional linear $Q1$-element with small deformations, plain stress and St. Venant-Kirchhoff material:

$$\Pi_{Q1}^{\text{int}} = \frac{1}{2} \int_{-\frac{b}{2}}^{\frac{b}{2}} \int_{-\frac{a}{2}}^{\frac{a}{2}} \alpha(\sigma_{\xi\xi}\varepsilon_{\xi\xi} + \sigma_{\xi\eta}\varepsilon_{\xi\eta}) \mathrm{d}\xi \mathrm{d}\eta = \alpha \frac{Eb^3 a}{24(1-\nu^2)} \left(1 + \frac{1-\nu}{2} \left(\frac{a}{b}\right)^2 \right) \quad (3.41)$$

with α scaling the deformation amount. It is clearly seen that the shear energy is scaled with the square of the aspect ratio $\frac{a}{b}$ and therefore dominates the energy for large ratios. This excessive internal shear energy spuriously balances the external energy underestimating the deformation. The element "locks" in bending and the driving critical parameter is identified as the aspect ratio $\frac{a}{b}$.

Again, this deficit is extended to three dimensions in a straightforward manner. There are several possibilities to remedy this locking phenomenon, such as (selective) under-integration, the *Assumed Natural Strain* (ANS) method and *Discrete Strain Gap* (DSG) method, the incompatible-modes approach and the EAS-method, as discussed in the following sections. It is noted that this effect has a strong similarity to *transverse shear locking* in structural elements like beam-, plate-, or shell-elements. Usually, successful techniques in two dimensions can be transferred to three-dimensional elements.

3.2.3 Trapezoidal Locking

The phenomenon of trapezoidal locking is also a geometric locking type where the driving parameter is the element aspect ratio (slenderness). It arises if *curved* structures are modeled. In this case a low-order discretization automatically leads to trapezoidal-shaped elements, as seen in Fig. 3.6 on the right. It has to be distinguished from the situation on the left where the elements' trapezoidal shape is a result of mesh distortion. We introduce the so-called director which points from the lower to the upper node. Then in the former case the director is perpendicular to the continuous structure, but not to the element reference plane (dashed line in Fig. 3.6), whereas in the latter case it is not perpendicular to neither one.

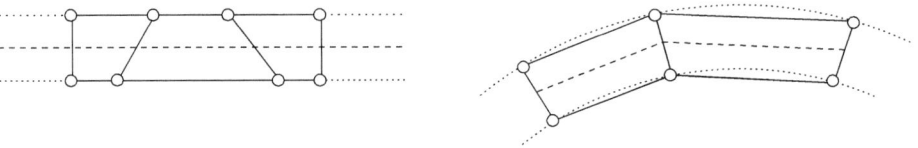

Fig. 3.6: Two different cases leading to trapezoidal shaped elements, where we distinguish element distortion (left) from a natural discretization of curved structure (right).

The reason for the locking effect is illustrated in Fig. 3.7. Here, a curved shell-like structure at a pure bending deformation state is discretized with three-dimensional solid elements. We consider one element, depicted in Fig. 3.7 (dashed) which clearly represents the characteristic trapezoidal shape. Also the deformed element (solid) and the discrete nodal bending forces acting in the shell-plane are sketched. It is seen that the discrete deformed shape automatically comes along with a shortening of the thickness between the nodes. The resulting strain $\varepsilon_{\zeta\zeta}$ is plotted in Fig. 3.7 below. In contrast, the thickness strain in the corresponding continuum problem is constantly equal to zero. This parasitic strain in the discrete element leads to spurious strain energy and therefore an underestimated deformation. The situation deteriorates with larger deviations of the director from the normal of the shell-plane which happens either for a larger slenderness of the element (side length/thickness) or for a growing curvature of the structure to be discretized.

The problem of trapezoidal shaped elements has been thoroughly examined in the late 80's by MacNeal [203, 205], though the distinction between mesh distortion and trapezoidal

3. Efficient Finite Elements for Arterial Wall Modeling

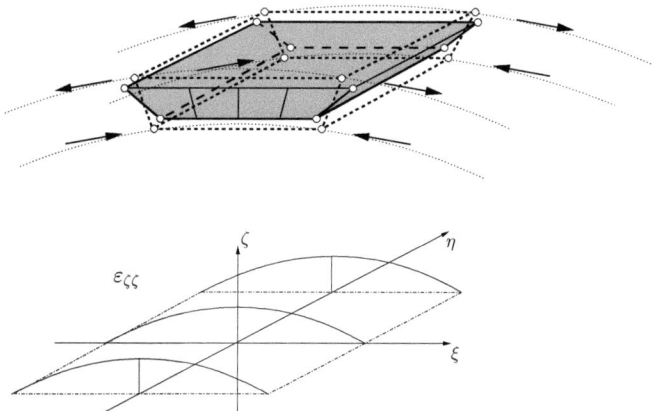

Fig. 3.7: Illustration of trapezoidal locking (above) with resulting thickness strain (below).

locking was not yet clear. The same effect, termed *curvature-thickness locking*, was discovered for structural shell-elements by Ramm *et al.* [242], see also Betsch *et al.* [30]. It gained importance when the trend went on to employing three-dimensional solid elements for shell-like structures. Trapezoidal locking can be eliminated by the ANS- and DSG-methods, but not by the EAS-method. We refer to Section 3.3.1 for implications.

3.2.4 Membrane Locking

Membrane locking is a phenomenon which is again related to the modeling of slender, shell-like structures in bending with three-dimensional solid elements. It arises only if the elements' faces are warped, that is their nodes do not lie within one plane. This typically happens only for doubly curved structures in certain cases. Fig. 3.8 depicts some example discretizations in which membrane locking does and does not occur. Also, the implications for linear elements are relatively small. We refer to the work of Koschnick *et al.* [175, 176] for a detailed treatise on membrane locking.

3.3 Limits of Element Perfectibility and the Issue of Mesh Quality

The goal of any finite element technique must be to obtain optimal results with reasonable effort for (relatively) coarse meshes, characterized by *coarse mesh accuracy*. As mentioned, many locking effects vanish for extreme fine meshes of the underlying structure where the element aspect ratio tends to one. However, this approach contradicts this goal and is not helpful in real problems.

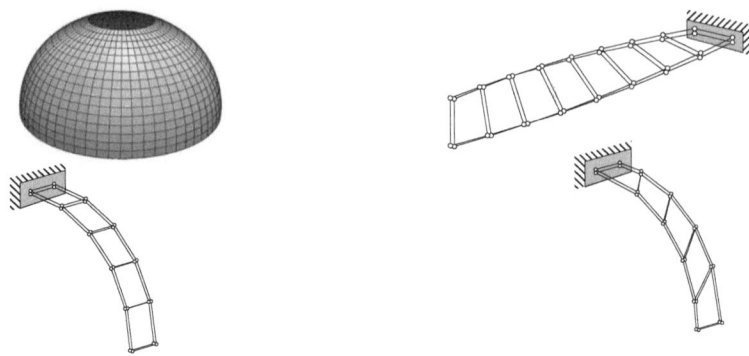

Fig. 3.8: Discretization examples without (left) and with (right) potential membrane locking.

But, there is no such thing as a perfect element for every application. There are always limitations on the effect and the efficiency of a certain element technology. As MacNeal [204] states:

> "The task of developing good finite elements never seems to be finished. Designers return, again and again, to the same basic configuration of nodes and find some way to eke out an improvement."

The mesh dependence is one such limitation which will be discussed in the following.

3.3.1 Element Perfectibility — MacNeal's Dilemma

The previous quote addresses an issue with linear elements in trapezoidal shape which is known as *MacNeal's Dilemma* [203] (see also [204, 205]). We have specified the problem of trapezoidal locking in Section 3.2.3. Another goal in finite element technology, besides eliminating locking, is obviously convergence to the correct result. Though neither necessary nor sufficient to prove convergence, the so-called *patch test*, first introduced by Irons [158], has always been a beneficial tool to assess element performance. It is based on the idea that a finite element should always be able to reproduce a constant strain state in every mesh configuration.

MacNeal's Dilemma now states that satisfaction of the patch-test and elimination of trapezoidal locking are *mutually exclusive* [203]. As soon as satisfaction of the constant strain patch test is assured, the element will exhibit trapezoidal locking. Let us consider the highlighted element of trapezoidal shape within the two discretizations illustrated in Fig. 3.9. In the case on the left the element has to represent a *constant* stress state, whereas in the case on the right it has to represent a *higher order* stress state due to bending. However, the element has the same geometric shape in both cases and the involved ansatz cannot distinguish the two cases. In fact, the correct results for each case are contrary and only one result may be represented exactly by the element.

3. Efficient Finite Elements for Arterial Wall Modeling

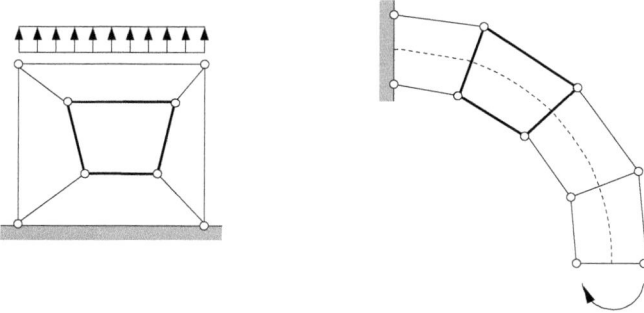

Fig. 3.9: Illustration of MacNeal's dilemma.

To handle this dilemma we propose the following (see also Frenzel *et al.* [97]): In the case of thick bulky structures where elements rarely show a tapered shape with high slenderness satisfaction of the patch test is given priority over elimination of trapezoidal locking. However, in situations where shell structures are modeled with solid elements trapezoidal locking is prominent and hence its elimination essential. Distorted meshes are likely to appear for practical applications, but fortunately the distortion is typically within the shell-plane. However, in the thickness direction the mesh layout is usually regular, resulting for example from extruding the shell-plane in the thickness direction as employed in Section Fig. 3.10. Mesh layouts which are distorted in thickness direction are very unlikely in practical applications. Therefore, a suitable three-dimensional solid-shell element should employ a technique for elimination of trapezoidal locking *in the thickness direction*, but ensure patch-test satisfaction *within the shell-plane*.

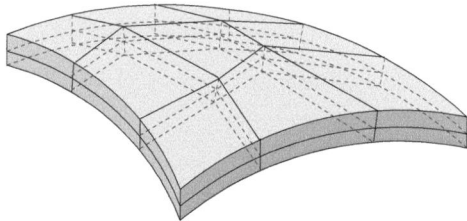

Fig. 3.10: Typical solid-shell discretization with in-plane mesh distortion, but regular out-of-plane mesh.

3.3.2 Influence of Mesh Quality

We have seen in the previous section that locking and mesh distortion are closely related. Although today's meshing algorithms seek to provide good quality meshes it can never be expected for complex structures that the element shapes are not distorted. Accuracy and convergence rate of finite elements strongly depend on the mesh distortion. It is worth mentioning that for a distorted element shape the Jacobian mapping is not polynomial anymore and thus cannot be evaluated exactly by the common Gauss quadrature. Unfortunately, distortion sensitivity appears to be much worse for advanced element technology, because all the ideas rely on an assumed strain or deformation state which becomes more complex in the distorted element.

This can be illustrated by the simple linear example in Fig. 3.11. The result of a two-element beam structure in bending is compared for a varying mesh distortion with different element technologies. Mind that the distortion is out-of-plane with respect to the bending load. On the right of Fig. 3.11 the result of the tip displacement, scaled by the exact value, is plotted against the varying distortion parameter a. The result of the standard displacement-based element Disp is due to shear locking significantly erroneous and the displacement is underestimated to about 20% for undistorted elements and 10% for extreme distortion. In contrast, all the other element techniques, described in detail in the following sections, give the exact result for the undistorted mesh. However, the performance drops for growing distortion, even below the standard element. Note that the sensitivity of the compared element techniques would differ significantly if the distortion would be in-plane with respect to the bending load.

This issue has to be taken into account when seeking coarse mesh accuracy, because there will always be some tradeoff with mesh quality. Obviously, it has always been the goal of research in finite element technology to optimize results with respect to locking, mesh dependence, and (computational) efficiency. However, fulfilling every demand seems to be impossible and no element technology comes without compromises.

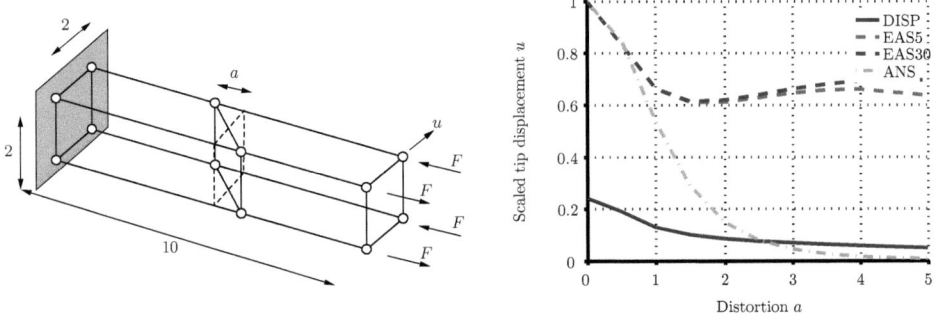

Fig. 3.11: Illustration of the influence of mesh distortion (out of plane) on different element techniques.

3.3.3 Locking in Large Deformation Analyses

The research in locking elimination techniques has mostly concentrated on small deformation analyses. Obviously, analytical results are easier obtained and the lack in accuracy due to locking can readily be observed within this limitation. Nevertheless, it is widely accepted to carry over the developed methods from small to large deformations and various elements are employed successfully. However, it is more difficult to single out specific effects of different locking influences in this case.

A number of stability issues have been observed for large deformation element technology. For example, an instability of EAS elements typically arises in problems involving homogeneous large compression, as reported by Armero [9], Wriggers and Reese [335], and Wall *et al.* [325]. Further stability implications on different advanced elements can be found at Harnau [123, 125] and Auricchio *et al.* [13] present a more rigorous stability study (see also Auricchio *et al.* [14]).

3.4 Methods to Eliminate Locking Phenomena

So far, the necessity of element technology was discussed. In this section some popular methods are elaborated, especially where they are applicable to three-dimensional elements. The specific element formulations are substantiated in Section 3.5.

3.4.1 Assumed (Natural) Strain Method

The idea of *assumed strains* has been pioneered by MacNeal [202] and Hughes and Tezduyar [151] who employed specific interpolations of transverse shear strains in plate- and shell-elements. The method has been systematically elaborated by Bathe and co-workers which has become the famous MITC-element family, see for instance the work of Bathe and Dvorkin [22, 23], or Bathe *et al.* [21]. Park and Stanley [226] utilize the strains at natural element coordinates, founding the name *Assumed Natural Strain* (ANS) method.

The principal idea of the ANS method is to choose a certain interpolation for the strains to be modified, instead of deriving them directly from the interpolation of the displacements. The modified interpolation is based on the strain evaluation at specific discrete *sampling points*. The selection of sampling points and interpolation determines the success of the method with respect to locking elimination. In Fig. 3.5 it is observed that the spurious shear strains are zero at the edge midpoints. If these locations are selected as sampling points, the correct zero shear can be interpolated within the element eliminating shear locking.

In the following presentation of the ANS method, exemplified for eliminating shear locking, we first consider the small deformation case in the discretized vector notation from Section 3.1.3. The preliminary assumption is the separation of the B-operator into shear- and normal-strain parts, $\mathbf{B}_s$ and $\mathbf{B}_n$, respectively, yielding

$$\mathbf{k}_{\text{lin}} = \int_{\mathcal{B}_0^{(e)}} \mathbf{B}^{\text{T}} \bar{\mathbb{C}} \, \mathbf{B} \, dV = \int_{\mathcal{B}_0^{(e)}} \mathbf{B}_n^{\text{T}} \bar{\mathbb{C}} \, \mathbf{B}_n \, dV + \int_{\mathcal{B}_0^{(e)}} \mathbf{B}_s^{\text{T}} \bar{\mathbb{C}} \, \mathbf{B}_s \, dV. \qquad (3.42)$$

The special interpolation of the modified strains, namely the shear strains, leads to a modified B-operator and thus to a modified element stiffness matrix

$$\mathbf{k}_{\text{lin}}^{\text{ANS}} = \int_{\mathcal{B}_0^{(e)}} \overline{\mathbf{B}}^{\text{T}} \bar{\mathbb{C}} \, \overline{\mathbf{B}} \, dV = \int_{\mathcal{B}_0^{(e)}} \mathbf{B}_n^{\text{T}} \bar{\mathbb{C}} \, \mathbf{B}_n \, dV + \int_{\mathcal{B}_0^{(e)}} \overline{\mathbf{B}_s}^{\text{T}} \bar{\mathbb{C}} \, \overline{\mathbf{B}_s} \, dV. \qquad (3.43)$$

The modified B-operator $\overline{\mathbf{B}_s}$ holds the chosen special interpolation of strains, therefore *assuming* the corresponding strain state. This modification of parts of the B-operator is a widespread technique also for other element technology methods, known as the so-called *B-bar methods*. For a discussion on the specific strain interpolation to eliminate locking we refer to Section 3.5.2 where we present an element formulation using the ANS method.

A variationally consistent proof of the method was presented afterwards by Simo and Hughes [269]. Because the strong form of the kinematic equation is violated by the modification of the strain interpolation, only the VHW principle can serve as variational basis. We repeat the VHW potential (3.4) for the small deformation case:

$$\Pi_{\text{VHW}}(\mathbf{u}, \boldsymbol{\varepsilon}, \boldsymbol{\sigma}) = \int_{\mathcal{B}_0} [\Psi(\boldsymbol{\varepsilon}) + \boldsymbol{\sigma} : (\boldsymbol{\varepsilon}^{\mathbf{u}} - \boldsymbol{\varepsilon}) - \varrho \mathbf{b} \cdot \mathbf{u}] \, dV + \Pi^{\text{ext}}. \tag{3.44}$$

The independent strain field $\boldsymbol{\varepsilon}$ is herein filled by the *assumed* strain field

$$\boldsymbol{\varepsilon} = \boldsymbol{\varepsilon}_n + \overline{\boldsymbol{\varepsilon}_s} \tag{3.45}$$

resulting from the previous modification and the separated B-operators. The independent stress field is eliminated by the orthogonality condition which states that the independent stress field $\boldsymbol{\sigma}$ is chosen to be orthogonal to the strain difference $\boldsymbol{\varepsilon}^{\mathbf{u}} - \boldsymbol{\varepsilon}$, yielding

$$\int_{\mathcal{B}_0} \boldsymbol{\sigma} : (\boldsymbol{\varepsilon}^{\mathbf{u}} - \boldsymbol{\varepsilon}) \, dV = 0. \tag{3.46}$$

Thus, only the assumed strain field contributes to the internal energy and the displacement-based strain field which causes the spurious locking energy to be eliminated.

The extension of the ANS method to large deformations based on the Green-Lagrange strains is straightforward. Accordingly, the Green-Lagrangean strains are assumed as

$$\mathbf{E} = \mathbf{E}_n + \overline{\mathbf{E}_s}. \tag{3.47}$$

For discretization and linearization of a specific ANS modification for large deformation finite elements we again refer to Section 3.5.2.

However, we want to remark that the derivation of the ANS modification is based on a modified interpolation of the *Green-Lagrange strains* $\mathbf{E}$, but not on the more general *deformation gradient* $\mathbf{F}$. We are not aware of any attempt in the literature to apply the ANS method to directly modify the deformation gradient. This is certainly a limitation of the ANS method as several material concepts, such as a multiplicative decomposition of the deformation gradient for plasticity, rely on this concept (see for example the work of Simo and colleagues [264, 265, 266, 270, 271]). One option to implement hyperelastoplasticity at finite strains together with ANS relies on the spectral decomposition of the Cauchy-Green tensor $\mathbf{C}$ (see Tan and Vu-Quoc [292] and references therein).

Another option relies on a multiplicative decomposition of the deformation gradient $\mathbf{F}$. To retain a deformation gradient $\mathbf{F}^{\text{mod}}$ which is consistent with a modified Green-Lagrange strain E^{mod} we follow the approach proposed by Hauptmann et al. [130], employed to the ANS modification (3.47). Generally, the deformation gradient can be decomposed into the right-stretch tensor U and an orthogonal rotation tensor R

$$F = RU. \tag{3.48}$$

If this is inserted into the Green-Lagrange tensor

$$E = \tfrac{1}{2}(F^{\text{T}}F - I) = \tfrac{1}{2}(U^{\text{T}}(R^{\text{T}}R)U - I) = \tfrac{1}{2}(U^2 - I), \tag{3.49}$$

the orthogonal rotation tensor R drops out and E depends solely on U. Via a polar decomposition of E^{mod} the modified right-stretch tensor U^{mod} is obtained

$$E^{\text{mod}} = \tfrac{1}{2}((U^{\text{mod}})^2 - I) \xrightarrow{\text{Polar Dec.}} U^{\text{mod}} \tag{3.50}$$

The sought modified deformation gradient F^{mod} can be evaluated as

$$F^{\text{mod}} = RU^{\text{mod}}, \tag{3.51}$$

where the unmodified rotation tensor R is obtained by a polar decomposition of equation (3.48). This procedure needs twice a polar decomposition and thus requires considerable computational effort.

However, the present work considers only hyperelastic materials, typically based on the Cauchy-Green tensor C which is readily available from E^{mod} as

$$C^{\text{mod}} = 2E^{\text{mod}} + I. \tag{3.52}$$

The evaluated stresses and elasticity matrices (see Chapter 4) are then free from spurious locking influences. The deformation gradient might be necessary to map stress quantities between configurations, for instance to evaluate Cauchy stresses from Second Piola-Kirchhoff stresses. For such a mapping of a consistent, locking-free stress quantity between material and current frames we suggest for efficiency to employ the standard displacement-based deformation gradient. We have not observed relevant differences compared to employing a consistent deformation gradient F^{mod}. However, this might be different in some special cases and a closer investigation is recommended.

3.4.2 The Discrete Strain Gap Method

A relatively new and less prevalent method to eliminate locking is the *DSG* method. Bletzinger et al. [40] have introduced this method named "Discrete Shear Gap method" to successfully eliminate transverse shear locking in beam-, plate-, and shell-elements which has been extended to large deformations by Bischoff [34]. As in other popular concepts like ANS and reduced integration, the constraint of zero shear deformation in a pure bending mode is relaxed by restricting the condition to discrete locations. The approach has been generalized, reflected also in renaming it to *Discrete Strain Gap* method, to eliminate other geometric locking phenomena such as membrane- and trapezoidal-locking by Koschnick et al. [176]. It turned out that it is a conceptually appealing approach to eliminate all kinds of geometric locking defects in elements independently of their shape and polynomial order. This includes structural elements like beam- or shell-elements as well as continuum elements. It has strong similarities to the ANS method and resulting element formulations are often identical.

This concept has been extended to the most general case of three-dimensional solid elements under large deformations by Frenzel et al. [99]. The formulation is briefly repeated below. The resulting 8-node hexahedral solid element is free from any kind of geometric locking, as demonstrated by Frenzel et al. [97, 98]. Within the present work we employ this method for the 6-node wedge-shaped solid-shell element, described in Section 3.5.3.

The basic idea of the DSG approach can be best explained by considering a classical shear-deformable (Timoshenko-) beam finite element suffering from transverse shear locking, which we however do not want to repeat here referring to the original paper of Bletzinger et al. [40]. The principal idea is to introduce a special integration and interpolation of the kinematic equations to be modified, for instance the transverse shear strains. The concept can be generalized, yielding a special integration rule of all strain components.

Following Koschnick et al. [176] (see also Frenzel et al. [97, 98]), we choose the following rule for the modified strain components expressed in local element coordinates (element parameter space):

$$E_{\xi\xi}^{\mathrm{DSG}} = \sum_{I=1}^{n_{nd}} N_{,\xi}^I \int_{\xi^1}^{\xi^I} \left(\mathbf{u}_{,\xi} \cdot \mathbf{G}_\xi + \tfrac{1}{2}(\mathbf{u}_{,\xi}^{\mathrm{T}} \cdot \mathbf{u}_{,\xi}) \right) \mathrm{d}\xi$$

$$E_{\eta\eta}^{\mathrm{DSG}} = \sum_{I=1}^{n_{nd}} N_{,\eta}^I \int_{\eta^1}^{\eta^I} \left(\mathbf{u}_{,\eta} \cdot \mathbf{G}_\eta + \tfrac{1}{2}(\mathbf{u}_{,\eta}^{\mathrm{T}} \cdot \mathbf{u}_{,\eta}) \right) \mathrm{d}\eta$$

$$E_{\zeta\zeta}^{\mathrm{DSG}} = \sum_{I=1}^{n_{nd}} N_{,\zeta}^I \int_{\zeta^1}^{\zeta^I} \left(\mathbf{u}_{,\zeta} \cdot \mathbf{G}_\zeta + \tfrac{1}{2}(\mathbf{u}_{,\zeta}^{\mathrm{T}} \cdot \mathbf{u}_{,\zeta}) \right) \mathrm{d}\zeta$$

$$E_{\xi\eta}^{\mathrm{DSG}} = \sum_{I=1}^{n_{nd}} N_{,\xi}^I \int_{\xi^1}^{\xi^I} \left(\sum_{J=1}^{n_{nd}} N_{,\eta}^J \int_{\eta^1}^{\eta^J} \tfrac{1}{2}\left(\mathbf{u}_{,\xi} \cdot \mathbf{G}_\eta + \mathbf{u}_{,\eta} \cdot \mathbf{G}_\xi + \mathbf{u}_{,\xi}^{\mathrm{T}} \cdot \mathbf{u}_{,\eta} \right) \mathrm{d}\eta \right) \mathrm{d}\xi$$

$$E_{\eta\zeta}^{\mathrm{DSG}} = \sum_{I=1}^{n_{nd}} N_{,\eta}^I \int_{\eta^1}^{\eta^I} \left(\sum_{J=1}^{n_{nd}} N_{,\zeta}^J \int_{\zeta^1}^{\zeta^J} \tfrac{1}{2}\left(\mathbf{u}_{,\eta} \cdot \mathbf{G}_\zeta + \mathbf{u}_{,\zeta} \cdot \mathbf{G}_\eta + \mathbf{u}_{,\eta}^{\mathrm{T}} \cdot \mathbf{u}_{,\zeta} \right) \mathrm{d}\zeta \right) \mathrm{d}\eta$$

$$E_{\xi\zeta}^{\mathrm{DSG}} = \sum_{I=1}^{n_{nd}} N_{,\xi}^I \int_{\xi^1}^{\xi^I} \left(\sum_{J=1}^{n_{nd}} N_{,\zeta}^J \int_{\zeta^1}^{\zeta^J} \tfrac{1}{2}\left(\mathbf{u}_{,\xi} \cdot \mathbf{G}_\zeta + \mathbf{u}_{,\zeta} \cdot \mathbf{G}_\xi + \mathbf{u}_{,\xi}^{\mathrm{T}} \cdot \mathbf{u}_{,\zeta} \right) \mathrm{d}\zeta \right) \mathrm{d}\xi. \quad (3.53)$$

Here we use the elemental covariant convective basis vectors $\mathbf{G}_\xi$, $\mathbf{G}_\eta$ and $\mathbf{G}_\zeta$ as defined in Section 3.5.2. It is seen that the strain components are first integrated from the reference (first) node to every other node. The integrals run along one parametric coordinate and are therefore one-dimensional. Note also that particular integrals are zero a priori. This scheme yields for every node a so-called *discrete strain gap*. These strain gaps are then interpolated within the element by the standard shape functions or more precisely by the corresponding derivatives because they concern strains. In case of shear strains this scheme is employed twice. The resulting modified strains $\boldsymbol{E}^{\mathrm{DSG}}(\xi,\eta,\zeta)$ are transformed to global Cartesian frame via standard transformation rules (see the following sections) yielding $\boldsymbol{E}^{\mathrm{DSG}}(x,y,z)$, which replace the discrete strains $\boldsymbol{E}$ of standard displacement based elements.

The close interrelation between DSG and ANS elements becomes clear by inspecting the above rule from a numerical point of view. The involved integrals are usually evaluated using numerical methods such as a standard Gauss rule. In the case of linear elements the required integrals are evaluated exactly employing one Gauss point at the center of the domain. Hence, the scheme of (3.53) with integration along the local coordinates to a specified node yields an evaluation point at the mid-point of the corresponding element edge. The same locations typically serve as sampling points in an ANS approach, for instance for a four-noded shell element. This match might be lost for higher order elements or other element shapes. The benefit of the DSG method is that no a priori definition of such sampling points is necessary, as these turn out naturally from the scheme in (3.53). Similar to the ANS method, the modification implies a modified B-operator identifying the DSG method as B-bar method. Thus, the VHW principle serves as a variational basis, though to our knowledge a rigorous mathematical analysis is still missing.

The DSG method shares another characteristic with the ANS method which is the failure of the constant strain patch test. This is not a problem, if the modification is not in the element plane as for classical shell and plate elements, but it is significant for the presented solid element. One attempt to overcome this issue is a formulation based on a decomposition of displacement-modes. This is a popular approach followed by several authors for a number of element formulations (see Section 3.4.4). Following Belytschko and Bindeman [27] we can write the ith component of the trilinear displacement field of the 8-node hexahedral element in terms of eight arbitrary constants a_{0i} to a_{3i}, and c_{1i} to c_{4i}, the spatial coordinates (X, Y, Z) and the parametric coordinates (ξ, η, ζ) as

$$u_i = a_{0i} + a_{1i}X + a_{2i}Y + a_{3i}Z + c_{1i}h_1 + c_{2i}h_2 + c_{3i}h_3 + c_{4i}h_4, \tag{3.54}$$

where

$$h_1 = \eta\zeta, \qquad h_2 = \xi\zeta, \qquad h_3 = \xi\eta, \qquad h_4 = \xi\eta\zeta. \tag{3.55}$$

This yields for the discrete finite element a set of eight equations for the nodal displacements:

$$\mathbf{d}_i = a_{0i}\mathbf{s} + a_{1i}\bar{\mathbf{X}}_1 + a_{2i}\bar{\mathbf{X}}_2 + a_{3i}\bar{\mathbf{X}}_3 + c_{1i}\mathbf{h}_1 + c_{2i}\mathbf{h}_2 + c_{3i}\mathbf{h}_3 + c_{4i}\mathbf{h}_4, \tag{3.56}$$

where $\bar{\mathbf{X}}_i$ are vectors consisting of the ith nodal coordinates and $\mathbf{s}$ and $\mathbf{h}_\alpha$ are defined as

$$\begin{aligned}
\mathbf{s}^\mathrm{T} &= (1, 1, 1, 1, 1, 1, 1, 1), \\
\mathbf{h}_1^\mathrm{T} &= (1, 1,\ -1,\ -1,\ -1,\ -1, 1, 1), \\
\mathbf{h}_2^\mathrm{T} &= (1,\ -1,\ -1, 1,\ -1, 1, 1,\ -1), \\
\mathbf{h}_3^\mathrm{T} &= (1,\ -1, 1,\ -1, 1,\ -1, 1,\ -1), \\
\mathbf{h}_4^\mathrm{T} &= (-1, 1,\ -1, 1, 1,\ -1, 1,\ -1).
\end{aligned} \qquad (3.57)$$

This definition represents a split of the elements displacement into one rigid body mode $\mathbf{s}$, three constant strain modes $\bar{\mathbf{X}}_1, \bar{\mathbf{X}}_2, \bar{\mathbf{X}}_3$, and four higher order modes $\mathbf{h}_1$ to $\mathbf{h}_4$. Fig. 3.12 sketches these modes for one coordinate direction.

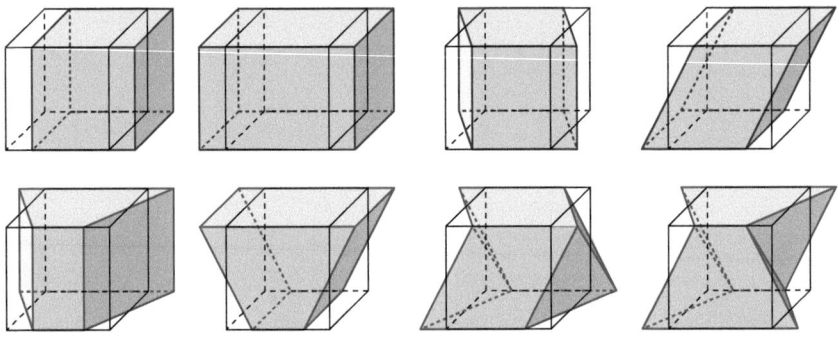

Fig. 3.12: Displacement modes of the 8-node hexahedral solid, illustrated for one arbitrary chosen coordinate direction: constant modes (first row, blue) and higher-order modes (second row, red).

It is straightforward to identify the constant strain modes as the ones assuring satisfaction of the constant-strain patch-test, whereas the higher-order modes are responsible for locking defects. It is therefore a goal in finite element design to deal only with the higher order modes to eliminate locking, but keep the other modes untouched to allow patch-test satisfaction. The key-component of this idea is the correct separation of these deformation modes. For instance, in two-dimensional 4-node elements this separation is easy to obtain, see among others Andelfinger [5] for details. However, for three-dimensional elements this is much more delicate. Where the constant modes are relatively simple to identify, for example via reduced integration, the higher-order modes seem to be not exactly separable for a general distorted element shape. We are not aware of any general solution of this issue in literature.

The idea, proposed earlier by Frenzel et al. [97, 98], that one could modify just the higher-order modes with for example the DSG approach to eliminate locking, fails for that reason, that the exact identification of these higher-order modes is not possible for an arbitrary distorted element shape. This is in accordance with the fact that also under-integrated elements are not perfect in any situation, because a successful stabilization for arbitrary distorted elements would also rely on the exact identification of these modes. Another link to this issue is the fact that the objective and variationally correct formulation of the popular EAS method similarly relies on the introduction of a (constant) metric defined at the element centre (see Section 3.5.1).

Finishing this section, we refer once more to Section 3.3 about element perfectibility, where the limitations of finite element design are already pointed out. The DSG-method has its merits in the conceptual beauty, but has also its shortcomings, most of them shared with the ANS method. We apply this method in the presented wedge-shaped element (see Section 3.5.3), but prefer the more popular ANS method in other element formulations, though the DSG-method would deliver the same results.

3.4.3 The Enhanced Assumed Strain Method

This method has been pioneered by Simo and Rifai [272] and has achieved a large impact in the field of element technology, reflected in the amount of literature about this topic. One advantage is that it is able to eliminate the volumetric locking effect caused by near incompressibility. Selected important references are the contributions of Simo and Armero [267], Simo *et al.* [268], and Andelfinger and Ramm [6].

The starting point of the *Enhanced Assumed Strain* (EAS) method is the VHW principle described in Section 3.1.1, however with the following reparametrization of the strains

$$\tilde{\boldsymbol{E}} := \boldsymbol{E} - \boldsymbol{E}^{\mathbf{u}} \tag{3.58}$$

$$\delta \tilde{\boldsymbol{E}} = \delta \boldsymbol{E} - \delta \boldsymbol{E}^{\mathbf{u}}. \tag{3.59}$$

This relates the additional strains $\tilde{\boldsymbol{E}}$ with the total strains $\boldsymbol{E}$ and the strains $\boldsymbol{E}^{\mathbf{u}}$ depending on the displacements. In this sense $\tilde{\boldsymbol{E}}$ could also be interpreted as the residuum of the kinematic equation $\boldsymbol{E} - \boldsymbol{E}^{\mathbf{u}} = 0$. Insertion into the VHW principle (3.4) yields

$$\Pi_{\text{EAS}}(\mathbf{u}, \tilde{\boldsymbol{E}}, \boldsymbol{S}) = \int_{\mathcal{B}_0} \Psi(\mathbf{u}, \tilde{\boldsymbol{E}}) \, dV - \int_{dV} \boldsymbol{S} : \tilde{\boldsymbol{E}} \, dV \\ - \int_{dV} \varrho \mathbf{b} \cdot \mathbf{u} \, dV - \int_{\Gamma_N} \hat{\mathbf{t}} \cdot \mathbf{u} \, dA + \int_{\Gamma_D} \mathbf{t}^S \cdot (\hat{\mathbf{u}} - \mathbf{u}) \, dA \quad \to \quad \text{stat.} \tag{3.60}$$

Variation with respect to the independent variables $\mathbf{u}, \tilde{\boldsymbol{E}}, \boldsymbol{S}$ reads

$$\delta\Pi_{\text{EAS}}(\mathbf{u}, \tilde{\boldsymbol{E}}, \boldsymbol{S}) = \int_{\mathcal{B}_0} \overbrace{\frac{\partial \Psi}{\partial \boldsymbol{E}^{\mathbf{u}}}}^{\boldsymbol{S}^{\mathbf{u}}} : \delta \boldsymbol{E}^{\mathbf{u}} \, dV + \int_{\mathcal{B}_0} \left(\overbrace{\frac{\partial \Psi}{\partial \tilde{\boldsymbol{E}}}}^{\boldsymbol{S}^{\tilde{E}}} - \boldsymbol{S} \right) : \delta \tilde{\boldsymbol{E}} \, dV \\ - \int_{\mathcal{B}_0} \delta \boldsymbol{S} : \tilde{\boldsymbol{E}} \, dV - \int_{\mathcal{B}_0} \varrho \mathbf{b} \cdot \delta \mathbf{u} \, dV \\ + \int_{\Gamma_D} (\hat{\mathbf{u}} - \mathbf{u}) \cdot \delta \mathbf{t}^S \, dA - \int_{\Gamma_N} (\hat{\mathbf{t}} - \mathbf{t}^{\mathbf{u}}) \cdot \delta \mathbf{u} \, dA = 0 \tag{3.61}$$

Using the fundamental lemma of variational calculus yields the local Euler-Lagrange equations:

$$\text{Div}(\boldsymbol{F}\boldsymbol{S}^u) + \varrho \boldsymbol{b} = \boldsymbol{0} \quad \text{in } \mathcal{B}_0 \qquad (3.62)$$

$$\boldsymbol{S}^{\tilde{E}} - \boldsymbol{S} = \boldsymbol{0} \quad \text{in } \mathcal{B}_0 \qquad (3.63)$$

$$\boldsymbol{E} - \boldsymbol{E}^u = \boldsymbol{0} \quad \text{in } \mathcal{B}_0 \qquad (3.64)$$

$$\hat{\boldsymbol{t}} - \boldsymbol{t}^u = \boldsymbol{0} \quad \text{on } \Gamma_N \qquad (3.65)$$

$$\hat{\boldsymbol{u}} - \boldsymbol{u} = \boldsymbol{0} \quad \text{on } \Gamma_D. \qquad (3.66)$$

The next step is the elimination of the stresses from the functional, again with the help of a certain orthogonality condition. If the independent stress field $\boldsymbol{S}$ is taken to be orthogonal to the enhanced strain field $\tilde{\boldsymbol{E}}$, which is equivalent to the vanishing integral

$$\int_{\mathcal{B}_0} \delta \boldsymbol{S} : \tilde{\boldsymbol{E}} \, \mathrm{d}V = 0 \quad \text{and} \quad \int_{\mathcal{B}_0} \boldsymbol{S} : \delta \tilde{\boldsymbol{E}} \, \mathrm{d}V = 0, \qquad (3.67)$$

the resulting formulation is considerably simplified:

$$\delta \Pi_{\text{EAS}}(\boldsymbol{u}, \tilde{\boldsymbol{E}}) = \int_{\mathcal{B}_0} \left[\frac{\partial \Psi}{\partial \boldsymbol{E}^u} : \delta \boldsymbol{E}^u + \frac{\partial \Psi}{\partial \tilde{\boldsymbol{E}}} : \delta \tilde{\boldsymbol{E}} - \varrho \boldsymbol{b} \cdot \delta \boldsymbol{u} \right] \mathrm{d}V$$

$$+ \int_{\Gamma_D} (\hat{\boldsymbol{u}} - \boldsymbol{u}) \cdot \delta \boldsymbol{t}^S \, \mathrm{d}A - \int_{\Gamma_N} (\hat{\boldsymbol{t}} - \boldsymbol{t}^u) \cdot \delta \boldsymbol{u} \, \mathrm{d}A = 0. \qquad (3.68)$$

It is worth mentioning that the orthogonality condition is usually only fulfilled after discretization. Linearization is given as

$$\mathrm{D}\,\delta\Pi_{\text{EAS}} \cdot (\Delta \boldsymbol{u}, \Delta \tilde{\boldsymbol{E}}) = \int_{\mathcal{B}_0} (\delta \boldsymbol{E}^u : \frac{\partial^2 \Psi}{\partial \boldsymbol{E}^2} : \Delta \boldsymbol{E}^u + \frac{\partial \Psi}{\partial \boldsymbol{E}} : \Delta \delta \boldsymbol{E}^u) \, \mathrm{d}V$$

$$+ \int_{\mathcal{B}_0} (\delta \boldsymbol{E}^u : \frac{\partial^2 \Psi}{\partial \boldsymbol{E}^2} : \Delta \tilde{\boldsymbol{E}} + \delta \tilde{\boldsymbol{E}} : \frac{\partial^2 \Psi}{\partial \boldsymbol{E}^2} : \Delta \boldsymbol{E}^u) \, \mathrm{d}V$$

$$+ \int_{\mathcal{B}_0} \delta \tilde{\boldsymbol{E}} : \frac{\partial^2 \Psi}{\partial \boldsymbol{E}^2} : \Delta \tilde{\boldsymbol{E}} \, \mathrm{d}V. \qquad (3.69)$$

We introduce the vector $\boldsymbol{\alpha}$ with variables α_i together with a matrix $\mathbf{M}$ to discretize on element level the enhanced strain $\tilde{\boldsymbol{E}}$, its variation and increment

$$\tilde{\boldsymbol{E}}^h = \mathbf{M}\,\boldsymbol{\alpha}, \qquad \delta \tilde{\boldsymbol{E}}^h = \mathbf{M}\,\delta\boldsymbol{\alpha}, \qquad \Delta \tilde{\boldsymbol{E}}^h = \mathbf{M}\,\Delta\boldsymbol{\alpha}. \qquad (3.70)$$

For convenience, we define on element level $\mathbf{S} := \frac{\partial \Psi}{\partial \boldsymbol{E}^u}$ and $\bar{\mathbb{C}} := \frac{\partial^2 \Psi}{\partial \boldsymbol{E}^{u2}}$ in vector and matrix notation, respectively (see Section 3.1.3). Further, we derive the discrete EAS-functional on element level yielding

$$\delta \Pi_{\text{EAS}}^{(e)} = \delta \mathbf{d}^T \underbrace{\int_{\mathcal{B}_0^{(e)}} \mathbf{B}^T \, \mathbf{S} \, \mathrm{d}V}_{\mathbf{f}_{\text{int}}} + \delta \boldsymbol{\alpha}^T \underbrace{\int_{\mathcal{B}_0^{(e)}} \mathbf{M}^T \, \mathbf{S} \, \mathrm{d}V}_{\mathbf{f}_{\text{EAS}}} + \delta \mathbf{d}^T \mathbf{f}_{\text{ext}} \qquad (3.71)$$

3. Efficient Finite Elements for Arterial Wall Modeling

Linearization of this nonlinear equations yields

$$\begin{aligned}
D(\delta\Pi_{\text{EAS}}^{(e)})(\mathbf{d}, \boldsymbol{\alpha}) \cdot (\Delta\mathbf{d}, \Delta\boldsymbol{\alpha}) &= \frac{\partial(\delta\Pi_{\text{EAS}}^{(e)})}{\partial \mathbf{d}} \cdot \Delta\mathbf{d} + \frac{\partial(\delta\Pi_{\text{EAS}}^{(e)})}{\partial \boldsymbol{\alpha}} \cdot \Delta\boldsymbol{\alpha} \\
&= \left[\delta\mathbf{d}^T \mathbf{k}_{uu} + \delta\boldsymbol{\alpha}^T \mathbf{k}_{\alpha u}\right] \cdot \Delta\mathbf{d} + \left[\delta\mathbf{d}^T \mathbf{k}_{u\alpha} + \delta\boldsymbol{\alpha}^T \mathbf{k}_{\alpha\alpha}\right] \cdot \Delta\boldsymbol{\alpha} \\
&= \delta\mathbf{d}^T \left[\mathbf{k}_{uu}\Delta\mathbf{d} + \mathbf{k}_{u\alpha}\Delta\boldsymbol{\alpha}\right] + \delta\boldsymbol{\alpha}^T \left[\mathbf{k}_{\alpha u}\Delta\mathbf{d} + \mathbf{k}_{\alpha\alpha}\Delta\boldsymbol{\alpha}\right]
\end{aligned} \quad (3.72)$$

with the definition of sub matrices:

$$\mathbf{k}_{uu} = \int_{\mathcal{B}_0^{(e)}} \mathbf{B}^T \bar{\mathbb{C}} \, \mathbf{B} \, dV + \mathbf{k}_{\text{geo}} \quad (3.73)$$

$$\mathbf{k}_{\alpha u} = \int_{\mathcal{B}_0^{(e)}} \mathbf{M}^T \bar{\mathbb{C}} \, \mathbf{B} \, dV \quad (3.74)$$

$$\mathbf{k}_{u\alpha} = \int_{\mathcal{B}_0^{(e)}} \mathbf{B}^T \bar{\mathbb{C}} \, \mathbf{M} \, dV = \mathbf{k}_{\alpha u}^T \quad (3.75)$$

$$\mathbf{k}_{\alpha\alpha} = \int_{\mathcal{B}_0^{(e)}} \mathbf{M}^T \bar{\mathbb{C}} \, \mathbf{M} \, dV. \quad (3.76)$$

The overall system of equations can be conveniently written in matrix form as

$$\begin{bmatrix} \mathbf{k}_{uu} & \mathbf{k}_{\alpha u}^T \\ \mathbf{k}_{\alpha u} & \mathbf{k}_{\alpha\alpha} \end{bmatrix} \begin{bmatrix} \Delta\mathbf{d} \\ \Delta\boldsymbol{\alpha} \end{bmatrix} = \begin{bmatrix} \mathbf{f}_{\text{ext}} - \mathbf{f}_{\text{int}} \\ -\mathbf{f}_{\text{EAS}} \end{bmatrix} \quad (3.77)$$

If the enhancing strain $\tilde{E}$ is chosen to be discontinuous across the element boundaries, it is possible to eliminate the EAS parameter increment $\Delta\boldsymbol{\alpha}$ at the element level by so-called *static condensation*. Solving for the increment $\Delta\boldsymbol{\alpha}$ using the second row of (3.77)

$$\Delta\boldsymbol{\alpha} = -\mathbf{k}_{\alpha\alpha}^{-1} \left(\mathbf{f}_{\text{EAS}} + \mathbf{k}_{\alpha u}\Delta\mathbf{d}\right) \quad (3.78)$$

then substituting (3.78) into the first row yields the element stiffness matrix

$$\mathbf{k}_{(e)}^{\text{EAS}} = \mathbf{k}_{uu} - \mathbf{k}_{\alpha u}^T \mathbf{k}_{\alpha\alpha}^{-1} \mathbf{k}_{\alpha u} \quad (3.79)$$

and the element residual force vector

$$\mathbf{r}_{(e)}^{\text{EAS}} = \mathbf{f}_{\text{ext}} - \mathbf{f}_{\text{int}} + \mathbf{k}_{\alpha u}^T \mathbf{k}_{\alpha\alpha}^{-1} \mathbf{f}_{\text{EAS}}. \quad (3.80)$$

The global system is retained from the standard assembly of the element matrices $\mathbf{k}_{(e)}^{\text{EAS}}$ and residual force vectors, as

$$\mathbf{K} = \bigwedge_{e=1}^{n_{ele}} \mathbf{k}_{(e)}^{\text{EAS}} \quad \text{and} \quad \mathbf{R} = \bigwedge_{e=1}^{n_{ele}} \mathbf{r}_{(e)}^{\text{EAS}}. \quad (3.81)$$

The fact that only the element matrix $\mathbf{k}_{(e)}^{\text{EAS}}$ changes allows an easy implementation of such an element into an existing framework and is undoubtedly a reason for its success. However, the inversion of the matrix $\mathbf{k}_{\alpha\alpha}$ represents a significant computational effort at element level, especially for larger vectors $\boldsymbol{\alpha}$.

For implementing an EAS element the only remaining thing is to substantiate the ansatz for the enhanced strains $\tilde{\boldsymbol{E}} = \mathbf{M}\boldsymbol{\alpha}$ where satisfaction of the orthogonality condition (3.67) is worth emphasizing. We refer to Sections 3.5.1 and 3.5.2 for details.

We finally remark that it is also possible within the EAS approach to enhance the deformation gradient $\boldsymbol{F}$, as presented among others in the original paper by Simo and Rifai [272]. This usually requires more enhancing parameters, but on the other hand the deformation gradient allows the most general interface to any material law. But because we will combine the EAS method with the ANS method we stick to the formulation based on the Green-Lagrange strain, see also the discussion in Section 3.5.2.

3.4.4 Alternative Methods

The field of finite element technology offers a considerable number of further methods and strategies to improve element performance. A complete coverage seems almost impossible and goes well beyond the scope of the present work. We only want to address some important alternative techniques in this section where they are in our personal opinion quite popular, especially in the field of biomechanical applications.

One of the first successfully employed techniques which is still present in some commercial finite element programs is the approach of "incompatible modes", originally proposed by Wilson et al. [331] and later improved by Taylor et al. [299]. By enhancing the two-dimensional 4-node element with certain incompatible displacement modes the behavior with respect to shear and volumetric locking has been greatly improved. However, it was not before the landmark paper of Simo and Rifai [272] that the variational consistency of such a technique was provided (see also Simo and Armero [267]). Today the close relation with the EAS method is established, among others by Bischoff and Romero [36]. Within this group of element formulations we also want to mention recent developments directly based on the VHW-functional such as the elements proposed by Kasper and Taylor [165, 166]. For a mathematical treatment on the correlation of these methods refer to the work of Wohlmuth and colleagues (Djoko et al. [67], Chavan et al. [54] and references therein).

Another widespread technique is based on under-integration and suitable hourglass-stabilization. As elaborated in Section 3.4.2, the element deformation state can be separated into constant modes and higher-order modes, also known as hourglass modes. By applying a reduced integration rule the constant modes can be identified and represented without locking defects. However, so-called zero-energy modes (see Irons and Ahmad [157]) arise leading to wrong results and instabilities of the solution. Therefore, certain hourglass stabilization techniques were introduced to prevent element instabilities as possible without reintroducing the unwanted locking defects. The group of Belytschko and colleagues have significantly contributed to this field of research, where we just cite the papers related to hexahedral elements by Flanagan and Belytschko [91] and Belytschko and Bindeman [27] and refer to references therein for details.

3. Efficient Finite Elements for Arterial Wall Modeling

Within an explicit time-integration framework, these techniques have gained broad success. However, reduced-integration elements, especially three-dimensional hexahedrons, are highly sensitive to mesh distortion (see Tan and Vu-Quoc [292]), which becomes obvious as a generally successful stabilization would necessitate the rigorous identification of the higher-order modes.

Felippa and colleagues followed the idea of displacement mode decomposition as well in their so-called "free formulation" (see Bergan and Felippa [29], Felippa and Bergan [84] and Felippa [83]). Another popular group of advanced finite elements are the so-called *hybrid stress* elements, originated by Pian and colleagues [232, 233, 234] which are based on the two-field *Hellinger-Reissner variational principle*. More recently the group of Sze has contributed to this topic (see [283] and references therein). Especially successful in biomechanical applications when dealing with incompressibility is the *mixed Jacobian-pressure formulation* of Simo et al. [275] (see also Simo and Taylor [274]). In summary, elements based on two-field or three-field variational principles are known as *mixed finite elements*. We do not want to go into further details about this large group of element formulations and refer to the literature, for instance the books of Hughes [147], Zienkiewicz and Taylor [341], Zienkiewicz et al. [342], and a recent publication of Nakshatrala et al. [217].

A final remark on alternative element formulations considers the development of efficient linear tetrahedral elements. Unfortunately, none of the above mentioned approaches is applicable to significantly improve their performance, as explained in more detail in Section 3.5.3. The reason is their completeness in the set of polynomial shape functions. These elements are characterized by poor performance in bending and incompressible situations. The approach of Taylor [298] is a mixed formulation based on a three-field variational principle and is in this sense closely related to the above methods. However, the tetrahedral element formulation lacks the benefit of static condensation and therefore introduces additional global unknowns. This also complicates coupling to other element-types. Because there seems to be no way out of the limiting constant stress and constant deformation gradient within one element, a promising strategy is to take more than one single element into account. This road has been followed by a number of researchers, among them Bonet and colleagues [43, 44], Dohrmann et al. [68], de Souza Neto, Pires and colleagues [219, 235], and the work of Thoutireddy et al. [300] and Guo et al. [119]. The idea has been combined with a stabilization technique by Puso and Solberg [236] and has recently been further improved for problems with Poisson's ratio other than zero, especially in the regime of near-incompressibility, by Gee et al. [111]. The approach, termed "nodal integration", might resolve the issue of tetrahedral meshes in bending dominated problems and — even more demanded — in (nearly) incompressible problems. Especially in biomechanical applications the complex geometry often relies on tetrahedral meshes for efficient meshing strategies. However, further research on such element formulations is necessary before a widespread application is recommended.

3.5 Efficient 3D Finite Element Formulations

In this section we describe in detail a selection of suitable finite elements for biomechanical simulations. They do all tackle the various locking phenomena but do also compromise about computational efficiency and element perfectibility. Insofar they substantiate the methods described in the previous section.

3.5.1 A Bulky Hexahedral Element

The first presented element is a 8-node hexahedral element which is well suited for modeling bulky compact structures. In this respect it differs from the subsequently described solid-shell element. Such an 8-node hexahedron represents certainly a classical three-dimensional finite element discretization. It is the natural extension of the 4-node quadrilateral element which has been one of the first finite elements ever and exhibits also a long history in element technology investigations.

We apply the EAS-method to remedy locking tackling especially volumetric and shear locking. The work is based on the paper by Andelfinger and Ramm [6] where different small-deformation EAS elements are compared and analyzed. Their extension to large deformations is straight-forward (see also Klinkel and Wagner [173]).

The standard linear 8-node isoparametric solid element employs linear shape functions for geometry and displacement, given as

$$\mathbf{X}^{h8} = \sum_{I=1}^{8} N_I(\xi,\eta,\zeta)\bar{\mathbf{X}}_I \quad \text{and} \quad \mathbf{u}^{h8} = \sum_{I=1}^{8} N_I(\xi,\eta,\zeta)\mathbf{d}_I, \tag{3.82}$$

where the trilinear shape functions are defined as

$$N_I(\xi,\eta,\zeta) = \frac{1}{8}(1+\xi_I\xi)(1+\eta_I\eta)(1+\zeta_I\zeta). \tag{3.83}$$

Now a specified formulation for the interpolation of the *enhanced strains* $\tilde{\boldsymbol{E}}$ is introduced. Remember the orthogonality condition

$$\int_{\mathcal{B}_0} \delta\boldsymbol{S}:\tilde{\boldsymbol{E}}\,dV = 0 \tag{3.84}$$

which has to be fulfilled to allow elimination of the stresses as primary fields in the VHW functional. If the enhanced strain field is introduced in the form

$$\tilde{\boldsymbol{E}} = \mathbf{M}\,\boldsymbol{\alpha} \tag{3.85}$$

with $\boldsymbol{\alpha}$ as vector of n_{EAS} internal strain parameters and the condition holds at least for an assumed constant stress within $\mathcal{B}_0$, it implies that

$$\int_{\mathcal{B}_0} \mathbf{M}\,\boldsymbol{\alpha}\,dV \stackrel{!}{=} 0. \tag{3.86}$$

Such a interpolation is based on the natural element coordinates and therefore depends on the metric within the element. But because this metric changes within the element for distorted shapes, the idea of Simo and Rifai [272] is to relate the EAS interpolation to the element centre, thus holding it constant within the element. Accordingly, the enhanced strain field is transformed between parametric and material space with an intermediate Jacobian matrix evaluated at the element center, yielding the two-point tensor

$$\tilde{E}_{kl} := \frac{\det \mathbf{J}}{\det \mathbf{J}_0} J_{0ki} \tilde{E}_{ij} J_{0lj}. \tag{3.87}$$

Here $\mathbf{J}$ denotes the general Jacobian matrix and $\mathbf{J}_0$ the Jacobian matrix evaluated at the parametric center of the element:

$$\mathbf{J}_0 = \begin{bmatrix} \frac{\partial X}{\partial \xi} & \frac{\partial Y}{\partial \xi} & \frac{\partial Z}{\partial \xi} \\ \frac{\partial X}{\partial \eta} & \frac{\partial Y}{\partial \eta} & \frac{\partial Z}{\partial \eta} \\ \frac{\partial X}{\partial \zeta} & \frac{\partial Y}{\partial \zeta} & \frac{\partial Z}{\partial \zeta} \end{bmatrix}_{\xi=0,\eta=0,\zeta=0}. \tag{3.88}$$

The definition of (3.87) maps the enhanced strains in material space $\tilde{\mathbf{E}}$ into the enhanced strains $\underline{\tilde{\mathbf{E}}}$ in parametric space. Using vector notation the inverse relation is given by

$$\underline{\tilde{\mathbf{E}}} = \frac{\det \mathbf{J}_0}{\det \mathbf{J}} \mathbf{T}_0^{-T} \underline{\tilde{\mathbf{E}}} \tag{3.89}$$

with the 6×6 matrix $\mathbf{T}_0$ defined as (see also Andelfinger and Ramm [6]):

$$\mathbf{T}_0 = \begin{bmatrix} J_{11}^2 & J_{21}^2 & J_{31}^2 & 2J_{11}J_{21} & 2J_{21}J_{31} & 2J_{11}J_{31} \\ J_{12}^2 & J_{22}^2 & J_{32}^2 & 2J_{12}J_{22} & 2J_{22}J_{32} & 2J_{12}J_{32} \\ J_{13}^2 & J_{23}^2 & J_{33}^2 & 2J_{13}J_{23} & 2J_{23}J_{33} & 2J_{13}J_{33} \\ J_{11}J_{12} & J_{21}J_{22} & J_{31}J_{32} & J_{11}J_{22}+J_{21}J_{12} & J_{21}J_{32}+J_{31}J_{22} & J_{11}J_{32}+J_{31}J_{12} \\ J_{12}J_{13} & J_{22}J_{23} & J_{32}J_{33} & J_{12}J_{23}+J_{22}J_{13} & J_{22}J_{33}+J_{32}J_{23} & J_{12}J_{33}+J_{32}J_{13} \\ J_{11}J_{13} & J_{21}J_{23} & J_{31}J_{33} & J_{11}J_{23}+J_{21}J_{13} & J_{21}J_{33}+J_{31}J_{23} & J_{11}J_{33}+J_{31}J_{13} \end{bmatrix}_0. \tag{3.90}$$

To introduce a suitable interpolation of $\underline{\tilde{\mathbf{E}}}$ the orthogonality condition, now obtained as

$$\int_{\mathcal{B}_0} \mathbf{M}\,\boldsymbol{\alpha}\,\mathrm{d}V = \int_{-1}^{1}\int_{-1}^{1}\int_{-1}^{1} \frac{\det \mathbf{J}_0}{\det \mathbf{J}} \mathbf{T}_0^{-T} \mathbf{M}(\xi,\eta,\zeta)\boldsymbol{\alpha}\,\mathrm{d}\xi\,\mathrm{d}\eta\,\mathrm{d}\zeta \stackrel{!}{=} 0, \tag{3.91}$$

simplifies the selection of appropriate polynomials for $\mathbf{M}(\xi,\eta,\zeta)$, as

$$\int_{-1}^{1}\int_{-1}^{1}\int_{-1}^{1} \mathbf{M}(\xi,\eta,\zeta)\,\mathrm{d}\xi\,\mathrm{d}\eta\,\mathrm{d}\zeta \stackrel{!}{=} 0 \tag{3.92}$$

must be satisfied.

Another consequence of condition (3.86) is that the interpolation of the enhanced strain field $\tilde{\mathbf{E}}$ is not contained in the displacement-based strain field $\mathbf{E}^u$. Furthermore, in case of a linear static theory it ensures stability and convergence of the resulting finite element method. In particular, satisfaction of the patch test is ensured a priori. For the large-deformation case such an interpolation also ensures frame invariance, a fundamental requirement of objectivity. We refer to Simo and Rifai [272] for details.

For our proposed 8-node solid elements we suggest two specific interpolations for $\mathbf{M}$, given in matrix notation. The first consists of nine enhanced strain parameters $n_{\mathrm{EAS}} = 9$ and is equivalent to an often employed incompatible displacement element using nine extra quadratic displacement modes. Its interpolation matrix is given as

$$\mathbf{M}^{\mathrm{EAS9}} = \begin{bmatrix} \xi & 0 & 0 & 0 & 0 & 0 & 0 & 0 & 0 \\ 0 & \eta & 0 & 0 & 0 & 0 & 0 & 0 & 0 \\ 0 & 0 & \zeta & 0 & 0 & 0 & 0 & 0 & 0 \\ 0 & 0 & 0 & \xi & \eta & 0 & 0 & 0 & 0 \\ 0 & 0 & 0 & 0 & 0 & \eta & \zeta & 0 & 0 \\ 0 & 0 & 0 & 0 & 0 & 0 & 0 & \xi & \zeta \end{bmatrix} \tag{3.93}$$

The resulting element, termed **EAS9**, is not competely free of volumetric locking in certain distorted element shapes (see Andelfinger and Ramm [6]). However, in terms of element perfectibility it is seen as good compromise between locking elimination and computational efficiency.

A second element, termed **EAS21**, consists of 21 enhancing strain parameters $n_{\mathrm{EAS}} = 21$ with the interpolation given as

$$\mathbf{M}^{\mathrm{EAS21}} = \begin{bmatrix} \xi & 0 & 0 & 0 & 0 & 0 & 0 & 0 & 0 & 0 & 0 & 0 \\ 0 & \eta & 0 & 0 & 0 & 0 & 0 & 0 & 0 & 0 & 0 & 0 \\ 0 & 0 & \zeta & 0 & 0 & 0 & 0 & 0 & 0 & 0 & 0 & 0 \\ 0 & 0 & 0 & \xi & \eta & 0 & 0 & 0 & 0 & 0 & 0 & 0 \\ 0 & 0 & 0 & 0 & 0 & \eta & \zeta & 0 & 0 & 0 & 0 & 0 \\ 0 & 0 & 0 & 0 & 0 & 0 & 0 & \xi & \zeta & 0 & 0 & 0 \\ 0 & 0 & 0 & 0 & 0 & 0 & \xi\eta & \xi\zeta & 0 & 0 & 0 & 0 \\ 0 & 0 & 0 & 0 & 0 & 0 & 0 & 0 & \xi\eta & \eta\zeta & 0 & 0 \\ 0 & 0 & 0 & 0 & 0 & 0 & 0 & 0 & 0 & 0 & \xi\zeta & \eta\zeta \\ \xi\zeta & \eta\zeta & 0 & 0 & 0 & 0 & 0 & 0 & 0 & 0 & 0 & 0 \\ 0 & 0 & \xi\eta & \xi\zeta & 0 & 0 & 0 & 0 & 0 & 0 & 0 & 0 \\ 0 & 0 & 0 & 0 & \xi\eta & \eta\zeta & 0 & 0 & 0 & 0 & 0 & 0 \end{bmatrix}. \tag{3.94}$$

Due to the overall nine introduced modes for the three normal strains, $\mathbf{E}_{\xi\xi}$, $\mathbf{E}_{\eta\eta}$, $\mathbf{E}_{\zeta\zeta}$, they all consist of the same polynomial basis and the constraint $\mathbf{E}_{ii} = 0$ can be fulfilled without an additional 'spurious' constraint. Thus, the element **EAS21** is free of volumetric locking. The additional six parameters for the shear strains eliminate shear locking in distorted element shapes. Additional nine modes ending up with 30 enhancing strain parameters would expand the whole strain field up to a complete trilinear field, as proposed by Klinkel and Wagner [173].

But this is considered to be computationally to expensive and according to Andelfinger and Ramm [6] does not seem to improve the overall performance.

Numerical examples are presented in Section 3.6 and the two proposed 8-node solid elements EAS9 and EAS21 prove to provide excellent results in these benchmark examples.

3.5.2 A Hexahedral Solid-Shell Element

In this section we present another 8-node hexahedral solid element. However, in contrast to the previous section this element is specialized to model thin, shell-like structures. These structures play a significant role in all fields of engineering which might be best described by Ramm's well-known quote characterizing the shell as the "primadonna of structures" [241]. For several reasons the trend in modeling shell structures goes to a three-dimensional discretization with special so-called *solid-shell* elements. Among them are the easy coupling to compact solid parts, the possible inclusion of higher-order effects such as boundary layers or delamination, and the direct identification of the discretized geometry, necessary for example in contact- or multifield-simulations. Not least, the arterial wall being a relatively thin, complex, layered structure is well suited to be modeled with solid-shell elements.

The notion 'solid-shell' is not completely clear in literature. A solid-shell typically consists of a three-dimensional discretization including a three-dimensional material law, but without any rotational degrees-of-freedom. However, much work is based on extensions of the classical two-dimensional shell theory into three dimensions, see for example Büchter *et al.* [26], Bischoff and Ramm [35], Betsch *et al.* [30], Parisch [223], Brank *et al.* [47], to name a few. This led some authors to reparametrize the shell geometry in terms of nodal displacements of upper and lower surfaces, see for example Schoop [257] and Schleebusch [256]. Other authors have taken the approach to use standard hexahedral elements and develop special techniques which allow the elements 'to become thin', for instance Legay and Combescure [192], or Key *et al.* [167]. The approach of Reese [243, 245] is based on a special stabilization technique for under-integrated 8-node solid elements resulting in computationally efficient solid-shell elements. However, stabilized under-integration is more worthwhile in an explicit dynamic framework and the element formulation is compromised by an inexact integration of the element volume for elements not representing a parallelepiped. In meshes resulting from complex patient-specific geometries of highly curved structures this might play a significant role.

Another widely accepted approach is represented by the work of Hauptmann and colleagues [130, 131, 132], see also Harnau and Schweizerhof [124, 125], and Klinkel *et al.* [170, 171], and Vu-Quoc and Tan [322, 323, 292]. These models have in common that they apply a certain combination of EAS- and ANS-techniques to overcome diverse locking defects. In contrast, Sze and colleagues [281, 283, 286, 287] mostly apply the hybrid-stress method in combination with ANS.

In the following we propose a 8-node solid-shell element as described by Vu-Quoc [322] which is also similar to the element proposed by Klinkel [170], but differs in the number of enhanced strain parameters. For a solid-shell element, as for the previous bulky hexahedral, we request to eliminate volumetric and shear locking. Within a shell geometry we may differentiate the latter into transverse and in-plane shear locking. Additionally, we have discussed the importance of elimination of trapezoidal locking in Section 3.2.3 which is preferred to patch-test satisfaction in thickness direction. Therefore, a distinguished thickness direction will be inherent in the present element formulation. We define without loss of generality the local ζ-direction as thickness direction.

Convective description for solid-shell elements

The classical shell theory is typically described using a convective local coordinate system and co- and contravariant basis vectors. The local finite element coordinate system also represents a convective description. We do not want to go into detail about these preliminaries and refer to various literature such as the books of Başar [15], Parisch [224], Itskov [159]. However, we do want to facilitate reference to the corresponding literature and give a short review.

To describe the solid-shell element geometry we introduce a local convective coordinate system where we can identify for every element point the generally curvilinear basis vectors. We distinguish between covariant (lower index) and contravariant (upper index) basis vectors. The covariant convective basis vectors are tangential to the curvilinear natural coordinates at a given point, yielding for the undeformed and deformed configuration, respectively,

$$\mathbf{G}_i = \frac{\partial \mathbf{X}}{\partial \xi_i} \quad \text{and} \quad \mathbf{g}_i = \frac{\partial \mathbf{x}}{\partial \xi_i}, \quad \text{with} \quad \mathbf{G}_i \cdot \mathbf{G}^j = \delta_i^j \quad \text{and} \quad \mathbf{g}_i \cdot \mathbf{g}^j = \delta_i^j. \tag{3.95}$$

To simplify notation the convective coordinates ξ_i are introduced which are equivalent to the local element parameter space coordinates ξ, η, ζ with $\xi_1 = \xi$, $\xi_2 = \eta$, $\xi_3 = \zeta$.

The covariant basis vectors are thus nothing more than the row vectors of the Jacobian matrix of the isoparametric map

$$\mathbf{J} = \begin{bmatrix} \frac{\partial X}{\partial \xi} & \frac{\partial Y}{\partial \xi} & \frac{\partial Z}{\partial \xi} \\ \frac{\partial X}{\partial \eta} & \frac{\partial Y}{\partial \eta} & \frac{\partial Z}{\partial \eta} \\ \frac{\partial X}{\partial \zeta} & \frac{\partial Y}{\partial \zeta} & \frac{\partial Z}{\partial \zeta} \end{bmatrix} = \begin{bmatrix} \mathbf{G}_1^T \\ \mathbf{G}_2^T \\ \mathbf{G}_3^T \end{bmatrix} \tag{3.96}$$

and the contravariant basis vectors can be identified as the column vectors of the inverse Jacobian

$$\begin{bmatrix} \mathbf{G}^1 & \mathbf{G}^2 & \mathbf{G}^3 \end{bmatrix} = \mathbf{J}^{-1} \tag{3.97}$$

The deformation gradient in global Cartesian coordinates

$$\boldsymbol{F} = \frac{\partial x_i}{\partial X_j} \mathbf{e}_i \otimes \mathbf{e}_j \tag{3.98}$$

is given in convective coordinates as

$$\boldsymbol{F} = \mathbf{g}_i \otimes \mathbf{G}^i \quad \text{and} \quad \boldsymbol{F}^{\mathrm{T}} = \mathbf{G}^i \otimes \mathbf{g}_i. \tag{3.99}$$

With the so-called metric coefficients of the current configuration $g_{ij} = \mathbf{g}_i \cdot \mathbf{g}_j$ and of the reference configuration $G_{ij} = \mathbf{G}_i \cdot \mathbf{G}_j$ the Green–Lagrange strain tensor reads

$$\mathbf{E} = E_{ij} \mathbf{G}^i \otimes \mathbf{G}^j \quad \text{with} \quad E_{ij} = \tfrac{1}{2}(g_{ij} - G_{ij}). \tag{3.100}$$

Strain modification stemming from for instance the ANS method are applied to these covariant strain components as described in the following.

ANS modification for the hexahedral solid-shell element

The discretization of the covariant basis vectors is given as

$$\mathbf{G}_i^h = \mathbf{N}_{,i}\,\bar{\mathbf{X}} \quad \text{and} \quad \mathbf{g}_i^h = \mathbf{N}_{,i}\,\bar{\mathbf{x}} \tag{3.101}$$

where $\mathbf{N}_{,i}$ is the derivative of $\mathbf{N}$ with respect to ξ_i and wherefrom the metric coefficients are evaluated. From here on the superscript h is omitted for readability. The Green-Lagrangean strain components (3.100) need to be transformed to global Cartesian coordinates. Similar to (3.90) we introduce a matrix $\mathbf{T}$ given as

$$\mathbf{T} = \begin{bmatrix}
J_{11}^2 & J_{21}^2 & J_{31}^2 & 2J_{11}J_{21} & 2J_{21}J_{31} & 2J_{11}J_{31} \\
J_{12}^2 & J_{22}^2 & J_{32}^2 & 2J_{12}J_{22} & 2J_{22}J_{32} & 2J_{12}J_{32} \\
J_{13}^2 & J_{23}^2 & J_{33}^2 & 2J_{13}J_{23} & 2J_{23}J_{33} & 2J_{13}J_{33} \\
J_{11}J_{12} & J_{21}J_{22} & J_{31}J_{32} & J_{11}J_{22}+J_{21}J_{12} & J_{21}J_{32}+J_{31}J_{22} & J_{11}J_{32}+J_{31}J_{12} \\
J_{12}J_{13} & J_{22}J_{23} & J_{32}J_{33} & J_{12}J_{23}+J_{22}J_{13} & J_{22}J_{33}+J_{32}J_{23} & J_{12}J_{33}+J_{32}J_{13} \\
J_{11}J_{13} & J_{21}J_{23} & J_{31}J_{33} & J_{11}J_{23}+J_{21}J_{13} & J_{21}J_{33}+J_{31}J_{23} & J_{11}J_{33}+J_{31}J_{13}
\end{bmatrix} \tag{3.102}$$

where in contrast to $\mathbf{T}_0$ of (3.90) the evaluation point is not the element center. The transformation of strains from local convective parameter space to global Cartesian space is obtained by premultiplying the transpose inverse matrix $\mathbf{T}^{-\mathrm{T}}$.

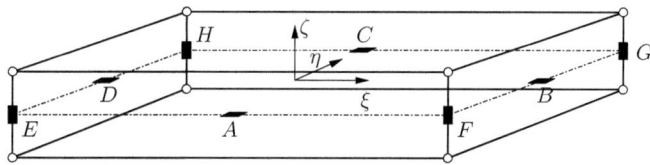

Fig. 3.13: 8-node solid shell element in isoparametric coordinates with sampling points for ANS interpolation of transverse shear strains (A, B, C, D) and transverse normal strains (E, F, G, H).

To treat transverse shear locking we adopt the ANS method, as first proposed by Dvorkin and Bathe [78] for a 4-node shell element. The compatible transverse shear strains $E_{\xi\zeta}^{\mathbf{u}}$ and $E_{\eta\zeta}^{\mathbf{u}}$ are evaluated at the four sampling points A, B, C, D, see Fig. 3.13, and the linear interpolation

$$E_{\xi\zeta}^{\mathrm{ANS}} = \tfrac{1}{2}(1-\eta)E_{\xi\zeta}^{\mathbf{u}}(\boldsymbol{\xi}_A) + \tfrac{1}{2}(1+\eta)E_{\xi\zeta}^{\mathbf{u}}(\boldsymbol{\xi}_C) \tag{3.103}$$

$$E_{\eta\zeta}^{\mathrm{ANS}} = \tfrac{1}{2}(1-\xi)E_{\eta\zeta}^{\mathbf{u}}(\boldsymbol{\xi}_D) + \tfrac{1}{2}(1+\xi)E_{\eta\zeta}^{\mathbf{u}}(\boldsymbol{\xi}_B) \tag{3.104}$$

is applied, where the coordinates of A, B, C, D are $\boldsymbol{\xi}_A = (0, -1, 0)$, $\boldsymbol{\xi}_B = (1, 0, 0)$, $\boldsymbol{\xi}_C = (0, 1, 0)$, $\boldsymbol{\xi}_D = (-1, 0, 0)$ in convective coordinates (ξ, η, ζ).

Trapezoidal locking is also tackled by ANS, as proposed by Betsch et al. [30]. Therefore, the compatible normal strain in the thickness direction $E_{\zeta\zeta}^{\mathbf{u}}$ is evaluated at the sampling points E, F, G, H, see Fig. 3.13, and bi-linearly interpolated within the element as

$$E_{\zeta\zeta}^{\text{ANS}} = \sum_{k \in \{E,F,G,H\}} N^k(\xi, \eta) E_{\zeta\zeta}^{\mathbf{u}}(\boldsymbol{\xi}_k) \qquad (3.105)$$

with $N^k = \frac{1}{4}(1 + \xi_k \xi)(1 + \eta_k \eta)$. The superscript indicates the shape function being evaluated at the designated location. The coordinates of the corner points E, F, G, H in convective coordinates (ξ, η, ζ) are $\boldsymbol{\xi}_E = (-1, -1, 0)$, $\boldsymbol{\xi}_F = (1, -1, 0)$, $\boldsymbol{\xi}_G = (1, 1, 0)$ and $\boldsymbol{\xi}_H = (-1, 1, 0)$.

Being a 'B-bar'-method the ANS-modification of natural strains is reflected in a modified B-operator matrix

$$\overline{\mathbf{B}}_I = \begin{bmatrix} N_{I,1} \cdot \mathbf{g}_1^T \\ N_{I,2} \cdot \mathbf{g}_2^T \\ \sum_{k \in \{E,F,G,H\}} \frac{1}{4}(1 + \xi_k \xi)(1 + \eta_k \eta) \, N_{I,3} \cdot \mathbf{g}_3^T \big|_k \\ N_{I,1} \cdot \mathbf{g}_2^T + N_{I,2} \cdot \mathbf{g}_1^T \\ \frac{1}{2}(1-\xi)(N_{I,2} \cdot \mathbf{g}_3^T \big|_{\boldsymbol{\xi}_D} + N_{I,3} \cdot \mathbf{g}_2^T \big|_{\boldsymbol{\xi}_D}) + \frac{1}{2}(1+\xi)(N_{I,2} \cdot \mathbf{g}_3^T \big|_{\boldsymbol{\xi}_B} + N_{I,3} \cdot \mathbf{g}_2^T \big|_{\boldsymbol{\xi}_B}) \\ \frac{1}{2}(1-\eta)(N_{I,1} \cdot \mathbf{g}_3^T \big|_{\boldsymbol{\xi}_A} + N_{I,3} \cdot \mathbf{g}_1^T \big|_{\boldsymbol{\xi}_A}) + \frac{1}{2}(1+\eta)(N_{I,1} \cdot \mathbf{g}_3^T \big|_{\boldsymbol{\xi}_C} + N_{I,3} \cdot \mathbf{g}_1^T \big|_{\boldsymbol{\xi}_C}) \end{bmatrix} \qquad (3.106)$$

for each nodal contribution I. The discrete virtual Green-Lagrangean strain-vector on element level is obtained from

$$\delta \mathbf{E} = \mathbf{T}^{-T} \overline{\mathbf{B}} \, \delta \mathbf{d} \quad \text{with} \quad \overline{\mathbf{B}} = [\overline{\mathbf{B}}_1 \,|\, \overline{\mathbf{B}}_2 \,|\, \overline{\mathbf{B}}_3 \,|\, \cdots \,|\, \overline{\mathbf{B}}_8]. \qquad (3.107)$$

The same interpolation is applied for the incremental Green-Lagrangean strain $\Delta \mathbf{E} = \overline{\mathbf{B}} \, \Delta \mathbf{d}$ and the modified element stiffness $\overline{\mathbf{k}}_{\text{eu}}$ then reads

$$\overline{\mathbf{k}}_{\text{eu}} = \overline{\mathbf{B}}^T \mathbf{T}^{-1} \bar{\mathbb{C}} \, \mathbf{T}^{-T} \overline{\mathbf{B}} \qquad (3.108)$$

The geometric matrix $\overline{\mathbf{k}}_{\text{geo}}$ resulting from linearization is derived in index notation referring to Equations (3.33) and (3.34), where the components $\hat{S}_{IK}$ are now modified to

$$\begin{aligned}
\overline{\hat{S}}_{IK} =\ & S_{11} N_{I,1} N_{J,1} + S_{22} N_{I,2} N_{J,2} \\
& + S_{33} \sum_{k=1}^{4} \tfrac{1}{4}(1 + \xi_k \xi)(1 + \eta_k \eta) N_{I,3}^k N_{J,3}^k \\
& + S_{12}(N_{I,1} N_{J,2} + N_{I,2} N_{J,1}) \\
& + S_{23} \left(\tfrac{1}{2}(1-\xi)(N_{I,2}^D N_{J,3}^D + N_{I,3}^D N_{J,2}^D) + \tfrac{1}{2}(1+\xi)(N_{I,2}^B N_{J,3}^B + N_{I,3}^B N_{J,2}^B) \right) \\
& + S_{13} \left(\tfrac{1}{2}(1-\eta)(N_{I,1}^A N_{J,3}^A + N_{I,3}^A N_{J,1}^A) + \tfrac{1}{2}(1+\eta)(N_{I,1}^C N_{J,3}^C + N_{I,3}^C N_{J,1}^C) \right).
\end{aligned}$$

$$(3.109)$$

EAS formulation for the hexahedral solid-shell element

In addition to the ANS modification we apply also the EAS method to treat volumetric locking, in-plane shear locking and membrane locking. The decisive question is the number and choice of enhancing strain parameters. We repeat Equation (3.91) here for convenience which specifies the enhancing EAS strains as

$$\tilde{\mathbf{E}} = \int_{-1}^{1}\int_{-1}^{1}\int_{-1}^{1} \frac{\det \mathbf{J}_0}{\det \mathbf{J}} \mathbf{T}_0^{-T} \mathbf{M}(\xi,\eta,\zeta) \boldsymbol{\alpha} \, d\xi \, d\eta \, d\zeta. \tag{3.110}$$

It remains to define the matrix $\mathbf{M}$ with respect to locking elimination and computational efficiency. Obviously, to eliminate volumetric locking the linear enhancement of the normal strains is necessary. Further, in-plane shear locking has not yet been eliminated by the ANS-method. The reason is that we anticipate irregular meshes within the shell plane and demand satisfaction of the in-plane membrane patch test which assures that a constant in-plane strain state can be reproduced exactly. This can be achieved by the EAS method through two additional linear enhancements of the in-plane shear strain $E_{\xi\eta}$. Finally, two bilinear polynomials for the thickness strain $E_{\zeta\zeta}$ are necessary to fully eliminate volumetric locking in bending and for distorted meshes, and thus to pass the out-of-plane bending patch test, suggested by MacNeal and Harder [206]. Vu-Quoc and Tan [322, 323] present an elaborate discussion about this issue.

The resulting enhancement matrix reads

$$\mathbf{M}^{\text{SOSH8}} = \begin{bmatrix} \xi & 0 & 0 & 0 & 0 & 0 & 0 \\ 0 & \eta & 0 & 0 & 0 & 0 & 0 \\ 0 & 0 & \zeta & 0 & 0 & \xi\zeta & \eta\zeta \\ 0 & 0 & 0 & \xi & \eta & 0 & 0 \\ 0 & 0 & 0 & 0 & 0 & 0 & 0 \\ 0 & 0 & 0 & 0 & 0 & 0 & 0 \end{bmatrix} \tag{3.111}$$

with $n_{\text{EAS}} = 7$ parameters $\alpha_1, \ldots, \alpha_7$. It is remarked that the in any case marginal effect of membrane-locking for linear elements is also eliminated by the enhancement of the membrane strains $E_{\xi\xi}$ and $E_{\eta\eta}$.

The overall Green-Lagrangean strain vector for the proposed solid-shell element is then given as the sum of the ANS-modified strains and the EAS-enhanced strains:

$$\mathbf{E}^{\text{SOSH8}} = \mathbf{T}^T \overline{\mathbf{B}} \, \mathbf{d} + \frac{\det \mathbf{J}_0}{\det \mathbf{J}} \mathbf{T}_0^{-T} \mathbf{M}^{\text{SOSH8}} \boldsymbol{\alpha}. \tag{3.112}$$

The resulting stiffness matrix is obtained as

$$\mathbf{k}^{\text{SOSH8}} = \underbrace{\overline{\mathbf{k}}_{uu}}_{\overline{\mathbf{k}}_{eu}+\overline{\mathbf{k}}_{geo}} - \overline{\mathbf{k}}_{\alpha u}^T \mathbf{k}_{\alpha\alpha}^{-1} \overline{\mathbf{k}}_{\alpha u} \tag{3.113}$$

The ANS modification affects only these parts of the stiffness matrix $\mathbf{k}_{uu}$ and $\mathbf{k}_{\alpha u}$ which are coupled to the nodal displacements, and the internal force vector $\mathbf{f}_{\text{int}}$, but not $\mathbf{k}_{\alpha\alpha}$ and $\mathbf{f}_{\text{EAS}}$.

This makes the implementation of the combined ANS and EAS modifications relatively simple and efficient.

In contrast, an EAS-enhancement of the deformation gradient $\boldsymbol{F} = \boldsymbol{F}^{\mathrm{u}} + \tilde{\boldsymbol{F}}$, as performed for example by Miehe [214], would result in a much more complicated formulation. The reason is that an EAS formulation based on the deformation gradient yields a multiplicative coupling between $\boldsymbol{F}^{\mathrm{u}}$ and $\tilde{\boldsymbol{F}}$ in $\boldsymbol{E}$. Therefore, a ANS modification would affect also parts of the stiffness matrix which are not directly dependent on the nodal displacements like for instance $\mathbf{k}_{\alpha\alpha}$. Such an approach is considered to be much less efficient (see also the discussions of Vu-Quoc [323] and Klinkel [169]). By all means the ANS modification cannot be applied directly to the deformation gradient and a material interface needs to be based on other strain measures (see Tan and Vu-Quoc [292] or Miehe [214]). Alternatively, we favor the polar decomposition approach described in Section 3.4.1.

In summary, the proposed solid-shell formulation is locking-free in the sense that it eliminates transverse shear locking, membrane locking, trapezoidal locking, and volumetric locking. It passes both the membrane and the out-of-plane bending patch tests. The only limitation is that the underlying geometry has to have a regular shell-discretization mesh-layout where the vector of the nodes in thickness direction is normal to the mid-plane. This is usually the case, especially if the mesh results from a kind of extrusion of a shell plane. We refer to Section 3.6 for several benchmark examples which prove the elements performance.

3.5.3 A Wedge-Shaped Solid-Shell Element

Complex geometries, especially coming from segmented patient-specific medical images, place high demands in terms of meshing algorithms and resulting mesh quality. Pure hexahedral meshes are practically impossible and hexa-dominant meshes are usually the best option. If the arterial wall is generated by extrusion of such meshes, see Section 6.1.2, also thin wedge-shaped elements emerge. This has motivated the development of a wedge-shaped 6-node solid-shell element, described in this section.

The problem of wedge-shaped elements lies in their triangular surfaces. It is well known that any linear triangular finite element has performance problems, whether it is a two-dimensional plane element, a triangular plate element, or a triangular shell element. This might be best characterized by Haußer and Ramm's contribution entitled, "Efficient 3-node shear deformable plate/shell elements — An almost hopeless undertaking"[129]. In fact, the linear triangle was one of the first finite elements and an enormous research effort has since been invested to improve it. Nevertheless, there still is not any fully satisfactory solution available (see also Haußer [128]).

Triangular elements are based on a complete set of polynomials, as illustrated by Pascal's triangle in Fig. 3.14 on the left. This is a key reason why triangular elements are barely improvable. In contrast, quadrilateral elements feature a mixed monomial and methods such as the EAS-method tackle locking problems at exactly this point by completing the polynomial set designated by this mixed monomial. This is not possible for triangles as they are already complete and would not "sense" the enhancing ansatz.

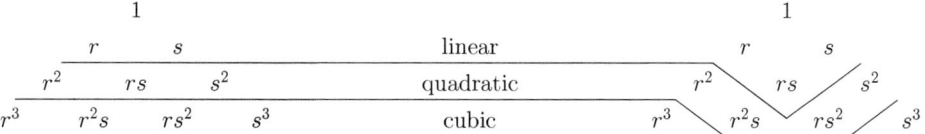

Fig. 3.14: Shape functions for triangular elements (left) and quadrilateral elements (right) in Pascal's triangle.

Displacement-based linear wedge-shaped solid element

Literature about wedge-shaped elements is rare. Only Sze and colleagues [282, 284] seem to have addressed these elements so far. However, when designing a wedge-shaped solid-shell element we can partly follow the vast literature on triangular shell elements (see for instance the early work of Hughes and Taylor [149] and Bletzinger et al. 's paper [40]). Further references and an in-depth elaboration are provided by Haußer [128] and Bischoff [34].

Prior to special treatment to improve its performance, we introduce the standard displacement-based wedge-shaped solid element. Depicted in Fig. 3.15, it is defined by six nodes and the shape functions resulting from a linear extension in the thickness direction of the linear triangle, yielding

$$N_1(r,s,\zeta) = \frac{1}{2}(1-\zeta)(1-r-s) \qquad N_4(r,s,\zeta) = \frac{1}{2}(1+\zeta)(1-r-s)$$
$$N_2(r,s,\zeta) = \frac{1}{2}(1-\zeta)r \qquad N_5(r,s,\zeta) = \frac{1}{2}(1+\zeta)r$$
$$N_3(r,s,\zeta) = \frac{1}{2}(1-\zeta)s \qquad N_6(r,s,\zeta) = \frac{1}{2}(1+\zeta)s. \qquad (3.114)$$

Similar to the hexahedral solid-shell element described in the previous section, we combine two different methods to best overcome locking phenomena. More specifically, the ANS/DSG method is applied to alleviate transverse shear locking and curvature thickness locking and the EAS method is employed to enhance the thickness strain.

ANS/DSG formulation for the wedge-shaped solid-shell element

As already discussed for the Sosh8-element, the ANS method is capable to eliminate trapezoidal locking (curvature thickness locking for shells) and is therefore employed for the wedge-shaped solid-shell as well. Accordingly, the transverse thickness strains $E_{\zeta\zeta}$ are modified. Note that, in contrast to the hexahedral element, the wedge-shaped solid-shell has already a predefined thickness direction, namely perpendicular to the triangular plane. We choose the corner points of the element C, D, E (see Fig. 3.15) as sampling points where the compatible strain $E_{\zeta\zeta}^{\text{u}}$ is

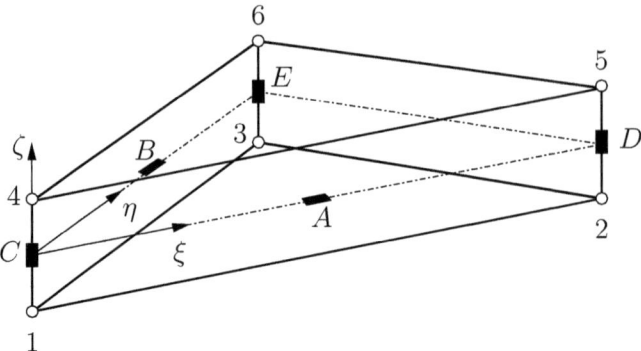

Fig. 3.15: 6-node wedge-shaped solid shell element in isoparametric coordinates with sampling points for ANS interpolation of transverse shear strains (A, B) and transverse normal strains (C, D, E).

evaluated. These strains are interpolated bi-linearly within the triangular element plane, given as

$$E_{\zeta\zeta}^{\text{ANS}} = \sum_{k \in \{C,D,E\}} N^k(r,s) E_{\zeta\zeta}^{\mathbf{u}}(\boldsymbol{\xi}_k), \quad (3.115)$$

with $N_1^{\text{tri}}(r,s) = r$, $N_2^{\text{tri}}(r,s) = s$, $N_3^{\text{tri}}(r,s) = 1 - r - s$ and the coordinates of the corner points C, D, E being $\boldsymbol{\xi}_C = (0,0,0)$, $\boldsymbol{\xi}_D = (1,0,0)$ and $\boldsymbol{\xi}_E = (0,1,0)$ in convective (parameter) coordinates (r, s, ζ).

We have already pointed out that there is a strong connection between the ANS and the DSG method. Typically the sampling points of the ANS method turn out to be the edge-integration points of the DSG approach, demonstrated as follows: Applying the DSG method to eliminate trapezoidal locking, the transverse strain $E_{\zeta\zeta}$ is taken as discrete strain gaps resulting from integration along the vertical edges and subsequently interpolated within the triangular element plane as

$$\begin{aligned}
E_{\zeta\zeta}^{\text{DSG}} &= \sum_{I=1}^{6} N_{,\zeta}^{I} \int_{\zeta^1}^{\zeta^I} \left(\mathbf{u}_{,\zeta} \cdot \mathbf{G}_\zeta + \tfrac{1}{2}(\mathbf{u}_{,\zeta}^{\mathrm{T}} \cdot \mathbf{u}_{,\zeta}) \right) \mathrm{d}\zeta \\
&= N_{,\zeta}^{4} \int_{\zeta^1}^{\zeta^4} \left(\mathbf{u}_{,\zeta} \cdot \mathbf{G}_\zeta + \tfrac{1}{2}(\mathbf{u}_{,\zeta}^{\mathrm{T}} \cdot \mathbf{u}_{,\zeta}) \right) \Big|_{r^4, s^4} \mathrm{d}\zeta \\
&\quad + N_{,\zeta}^{5} \int_{\zeta^1}^{\zeta^5} \left(\mathbf{u}_{,\zeta} \cdot \mathbf{G}_\zeta + \tfrac{1}{2}(\mathbf{u}_{,\zeta}^{\mathrm{T}} \cdot \mathbf{u}_{,\zeta}) \right) \Big|_{r^5, s^5} \mathrm{d}\zeta \\
&\quad + N_{,\zeta}^{6} \int_{\zeta^1}^{\zeta^6} \left(\mathbf{u}_{,\zeta} \cdot \mathbf{G}_\zeta + \tfrac{1}{2}(\mathbf{u}_{,\zeta}^{\mathrm{T}} \cdot \mathbf{u}_{,\zeta}) \right) \Big|_{r^6, s^6} \mathrm{d}\zeta.
\end{aligned} \quad (3.116)$$

By using a linear Gauss rule for the one-dimensional integration along ζ the evaluation points turn out to be exactly the corner points C, D, E. Obviously, the DSG approach yields the same strain modification as the ANS method specified in (3.115).

But, the situation changes for the transverse shear strains $E_{r\zeta}$ and $E_{s\zeta}$ and both methods significantly differ. Bletzinger et al. [40] have demonstrated a DSG approach to efficiently eliminate transverse shear locking for triangular shell elements. We have adopted this approach and propose the following DSG formulation for the wedge-shaped solid-shell element for large deformations. The modified shear strains read

$$E_{r\zeta}^{\text{DSG}} = \sum_{I=1}^{6} N_{,r}^{I} \int_{r^1}^{r^I} \left(\sum_{J=1}^{6} N_{,\zeta}^{J} \int_{\zeta^1}^{\zeta^J} \tfrac{1}{2} \left(\mathbf{u}_{,\zeta} \cdot \mathbf{G}_r + \mathbf{u}_{,r} \cdot \mathbf{G}_\zeta + \mathbf{u}_{,r}^{\text{T}} \cdot \mathbf{u}_{,\zeta} \right) \Big|_{r^J,s^J} \mathrm{d}\zeta \right) \Big|_{r^I,s^I} \mathrm{d}r, \qquad (3.117)$$

$$E_{s\zeta}^{\text{DSG}} = \sum_{I=1}^{6} N_{,s}^{I} \int_{s^1}^{s^I} \left(\sum_{J=1}^{6} N_{,\zeta}^{J} \int_{\zeta^1}^{\zeta^J} \tfrac{1}{2} \left(\mathbf{u}_{,\zeta} \cdot \mathbf{G}_s + \mathbf{u}_{,s} \cdot \mathbf{G}_\zeta + \mathbf{u}_{,s}^{\text{T}} \cdot \mathbf{u}_{,\zeta} \right) \Big|_{r^J,s^J} \mathrm{d}\zeta \right) \Big|_{r^I,s^I} \mathrm{d}s. \qquad (3.118)$$

The integration runs along the convective coordinates (r, s). Thus, thinking in the corresponding evaluation points (A and B in Fig. 3.15), only two edges are involved in this formulation. It turns out that the "discrete strain gaps" are rather an averaging along convective coordinates, than an averaging along the edges. This results in superior behavior with respect to locking, because the constraining third edge naturally drops out of the strain interpolation. In contrast, the proposed *Kirchhoff-Mode(KM)*-concept of Hughes and Taylor [149] for triangular shells represents an averaging along each of the three triangle edges and is still interfered by significant locking. An analogous ANS formulation could be easily formulated; however we are unaware of any literature reference proposing to take only two ANS sampling points into account. A very special ansatz proposed by Boisse [41] is mentioned, however the element is still not completely free from transverse shear locking (see Haußer [128] for details).

Following the idea of Koschnick [175] the performance of the triangular DSG shell element is optimal, if the local parameter space is oriented in such a way that the node is choosen as origin of the coordinates (r, s), whose included angle is closest to rectangular. This is also adopted for the proposed solid-shell wedge element. Obviously, such an element formulation is not objective with respect to node numbering and mesh topology. It is nevertheless shown in Section 3.6 that it shows excellent performance with respect to locking elimination.

EAS formulation for the wedge-shaped solid-shell element

For eliminating volumetric locking neither DSG nor ANS are capable and we employed the EAS method for the elements proposed up to now. Unfortunately, within the triangular plane of the wedge element the EAS method has no effect, because the polynomial set is complete. There is no simple and efficient method to improve the in-plane behavior available in literature. Maybe, the nodal-integration approach, see Section 3.4.4, could be applied, but an elaboration goes beyond the scope of this work.

Thus, the only remaining improvement can be achieved in thickness direction. Here, we employ one EAS enhancement parameter to remedy thickness locking. The corresponding enhancement matrix reads

$$\mathbf{M}^{\text{SOSH6}} = \begin{bmatrix} 0 \\ 0 \\ \zeta \\ 0 \\ 0 \\ 0 \end{bmatrix} \qquad (3.119)$$

Unfortunately, this does not completely eliminate thickness locking due to coupling of the in-plane and transverse normal stresses by Poisson's ratio. However, additional parameters such as $r\zeta$ and $s\zeta$ for $E_{\zeta\zeta}$ are not effective, again due to the complete polynomial set in triangular plane.

This is also the reason why the proposed wedge-shaped solid-shell element cannot pass the bending patch-test (see also Vu-Quoc [322]). Nevertheless, satisfaction of the membrane patch-test is ensured as only transverse strains are modified by ANS/DSG. The element is free from transverse shear-locking for an optimal mesh layout. Similar limitations are present in the wedge-shaped solid-shell element proposed by Sze et al. [284], which also necessitates a special element layout for optimal performance. Convergence, accuracy and efficiency are convincing, as demonstrated in the benchmark examples presented in the following section.

3.6 Numerical Benchmark Examples

In this section selected numerical examples are considered and the performance of different elements is analyzed. In the literature a large number of examples are available where more or less specific element technology features are compared. We present a small selection of popular examples to benchmark the performance of our proposed element set. Our results are always related to some literature results. However, a completely objective quantitative and qualitative measure of element performance, especially in the large deformation regime seems impossible. Rather, to specifically single out a certain element feature or locking characteristic, a parameter study is usually more illustrative. Furthermore, we want to point out that locking is present already in the small deformation regime and large deformations sometimes 'cover' the locking influence. However, all of our applications fall into the large deformation regime. Moreover, our proposed elements or at least the set of element technologies are already sufficiently discussed in the literature and therefore we do not repeat these examples here. The considered elements described in detail in the previous sections together with their labels are summarized in Table 3.1.

Table 3.1: Summary of considered elements for the following benchmark problems.

Identifier	Shape	Involved element technology	Cf. Sec.
EAS9	8-node hexahedron	9 EAS parameters	3.5.1
EAS21	8-node hexahedron	21 EAS parameters	3.5.1
Sosh8	8-node solid-shell	ANS, 7 EAS parameters	3.5.2
Sosh6	6-node solid-shell	ANS/DSG, 1 EAS parameter	3.5.3
$(\,\cdot\,)^{\mathrm{var}}$		specified variation of element $(\,\cdot\,)$	

3.6.1 Slit Annular Plate

This example is concerned with a slit annular plate clamped at one end and subject to a line load at the other end. It has been considered by several researchers, among them Büchter and Ramm [25], Wriggers and Gruttmann [334], Brank et al. [48]. Sze et al. [285] report detailed tabular load-displacement results which are taken as reference solutions. They are obtained by using 180 4-node shell elements with reduced integration and stabilization available in the commercial software ABAQUS.

Geometry
$R_a = 10.0$ m
$R_i = 6.0$ m
$h = 0.03$ m

Material
linear-elastic
$E = 2.1 \cdot 10^7$ kN/m^2
$\nu = 0.0$

Loading
$q = 26.67$ kN/m^2

Reference mesh
$6 \times 30 = 180$ elements

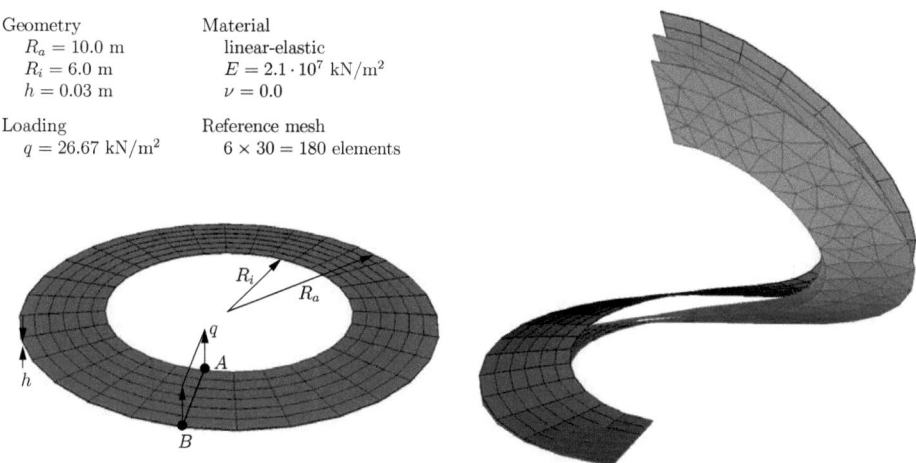

Fig. 3.16: Bending of slit annular plate with problem description (left) and three resulting displaced structures, corresponding to a hex-mesh (black) and two different wedge-meshes (red and green).

The geometry, boundary conditions and parameters are presented in Fig. 3.16 on the left. The final displaced structure is depicted in Fig. 3.16 on the right for three different discretizations. One is discretized with 180 hexahedral elements according to the reference (black outline) and the other two are discretized with wedge-shaped elements with different mesh layouts (red and green outline). The line load of 0.8 kN/m is transferred to the corresponding constant surface load of 26.67 kN/m^2 to adapt to the three-dimensional discretizations. The vertical tip displacement at point A and B are analyzed. To relate our results from a three-dimensional discretization to the reference specified at the mid-surface the top and bottom displacements are averaged. Poisson's ratio is taken to be zero and the structure is flat. Therefore only transverse shear locking is involved at the reference configuration, but the elements have a non-constant metric due to the circular structure.

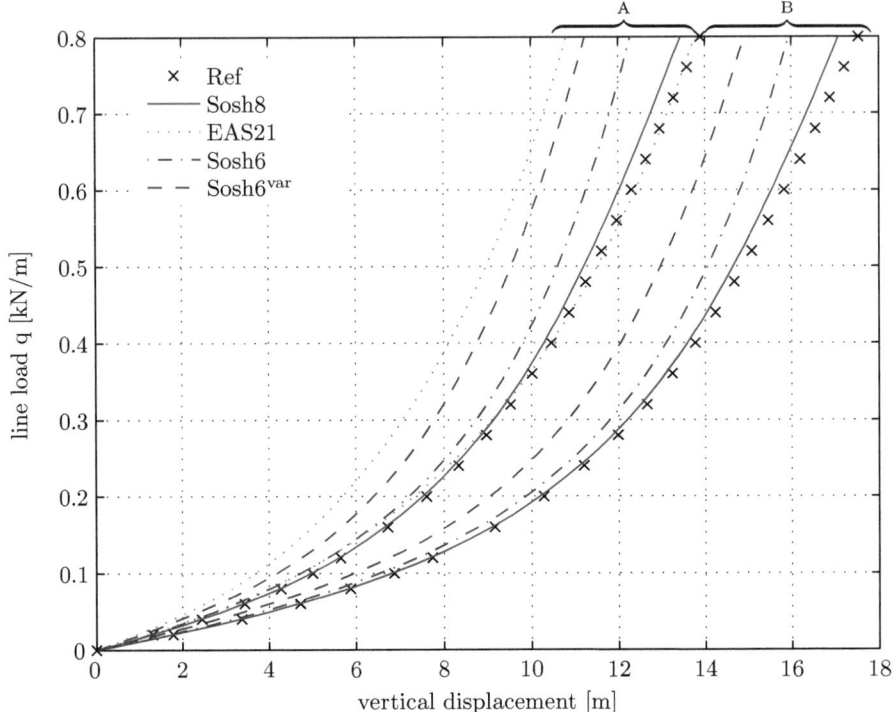

Fig. 3.17: Load-displacement diagram of the slit annular plate example.

In Fig. 3.17 the displacement of point A and B is plotted with respect to the line load of the shell discretization. The curves of the Sosh8-discretization are very close to the reference solution with only a slightly stiffer result. This is probably related to the difference between a shell-discretization and a fully three-dimensional discretization and the same results have been obtained by Hartmann [126]. In comparison, the results for the EAS21-discretization are considerably stiffer. This demonstrates the superior performance of ANS compared to EAS for the elimination of transverse shear-locking in such meshes where the elements have not the shape of a parallelogram.

Looking at the results of the wedge-shaped elements we consider two different mesh layouts. One is created by subdivision of every hexahedron of the reference discretization, see the red outlined mesh in Fig. 3.16. The other *variant* mesh results from a triangulation of the plate surface yielding 274 wedge elements. The displacements of the Sosh6-elements for the former discretizations are lower than the reference. However, taking into account the issues of triangular shell-elements (and wedge-shaped solid-shells) the results are remarkably close to the reference, especially at lower load levels. The DSG approach and the optimized orientation of the local triangular parameter space almost reach the performance of a 8-node solid-shell. In comparison, the results 'Sosh6^{var}' for the second mesh layout *without* the optimized parameter space orientation are significantly worse.

3.6.2 Pullout of Open-ended Circular Cylinder

Another popular example considers an open-ended circular cylinder subject to two pulling radial forces. The problem is presented by a number of researchers, among them Gruttmann et al. [117], Peng et al. [228], Brank et al. [48], Park et al. [225], Sansour and Kollmann [255]. As for the previous example Sze et al. [285] report detailed tabular results using the previously described shell element which we take as a reference solution. Geometry, material properties and loading conditions are given in Fig. 3.18 on the left. Owing to symmetry, one-eighth of the structure is modeled using 16 by 24 (384) hexahedral elements, applying appropriate symmetry boundary conditions. Another mesh is also depicted consisting of 858 wedge-shaped elements which results from the same nodal interval along the edges of the eighth structure. The displaced structure for the full load magnitude of 40 000 kN is depicted in Fig. 3.18 on the right.

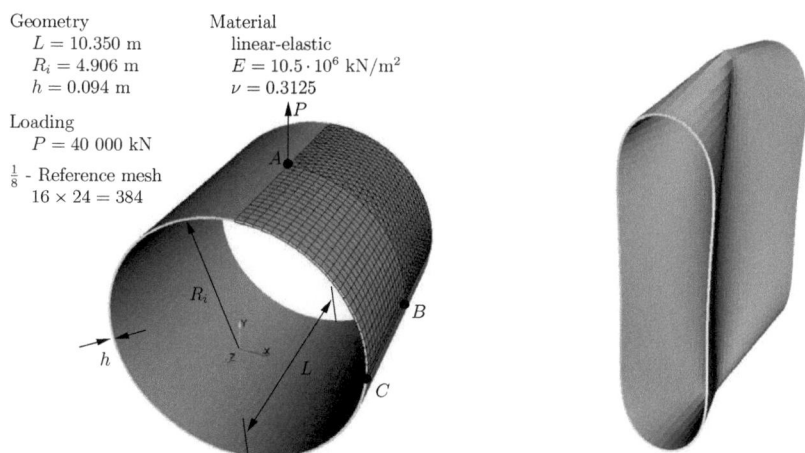

Fig. 3.18: Pullout of open-ended circular cylinder. Problem description with hex- and wedge-discretization (left) and resulting displaced structure (right).

In Fig. 3.19 several load-displacement diagrams are plotted considering three points A, B and C. Again, due to the three-dimensional discretizations we obtain the results by averaging nodal results from top and bottom surface. All curves reflect the characteristic displacement evolution owing to the buckling of the cylinder at about half the load magnitude. At this buckling load the displacements increase noticeable and at point C even the direction changes.

Comparing the results with the reference solution, close agreement can be observed for the Sosh8- and Sosh6-discretizations. The Sosh8 results are slightly softer compared to the reference result, which again likely results from the three-dimensional discretization. The Sosh6 solution is slightly stiffer although approximately twice as many elements are used. Still, the results are reasonably good. For comparison, we have also plotted the displacement curves for a wedge-shaped solid-shell element *without* the EAS parameter in the thickness-direction, labeled Sosh6^{var}. Clearly, the buckling load is underestimated in this case. With a relatively small aspect ratio of the individual elements the modeling of the correct thickness strain obviously plays a significant role and the EAS parameter successfully alleviates this deficit of the pure displacement formulation.

In this example, as well as in several other benchmark examples, a singular force and thus a singularity is involved. It should be noted that such examples do significantly depend on the resolution of the singularity by the surrounding mesh.

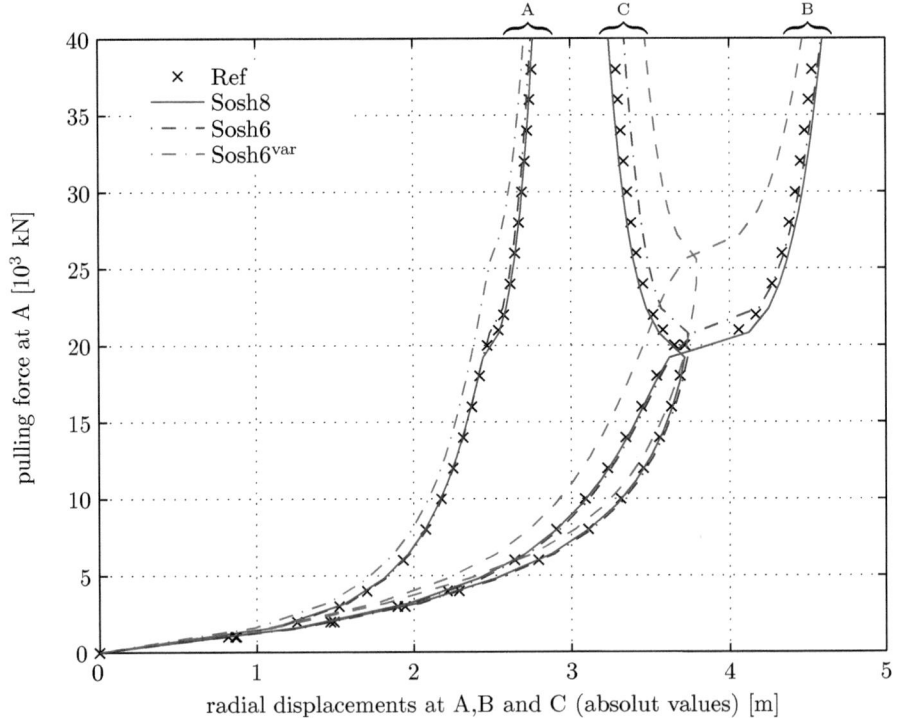

Fig. 3.19: Load-displacement diagram of open-ended circular cylinder.

3.6.3 Pinched Hemisphere

Despite the fact that it poses several issues on judging finite element results the pinched hemisphere example has become probably the most popular shell benchmark referenced in innumerable publications. It concerns a hollow hemispherical shell which has at its pole an 18° hole, as depicted on the left of Fig. 3.20. It is loaded by four alternating forces at the equator (compare the displaced structure in Fig. 3.20 on the right for illustration). Due to symmetry only a quarter of the shell is modeled and symmetry boundary conditions are applied accordingly. The structure itself is not fixed in the Z−direction which might pose a problem depending on the applied solver technique. We therefore fix one node at the middle of the equator in the vertical direction to eliminate rigid body motion. Note that two different thickness values of the structure are found in literature, namely $h = 0.04$ and $h = 0.01$. We consider the thin shell with $h = 0.01$, as presented among others by Simo et al. [273], Parisch [223], Betsch and Stein [31]. As reference we take the results from Klinkel et al. [170, 171].

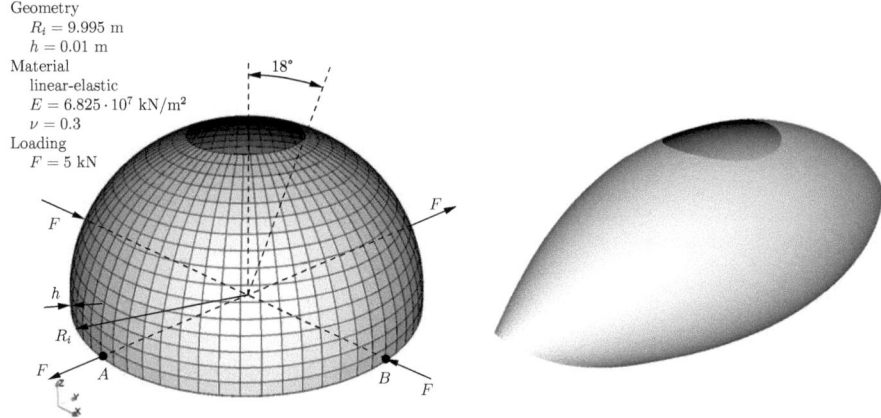

Fig. 3.20: Pinched hemisphere example with problem description (left) and resulting displaced structure (right).

In Fig. 3.21 the inward and outward displacements at points A and B are plotted versus the pinching load F. We compare our results when applying Sosh8-elements at three different mesh densities to the reference solution. As can be clearly observed our results for a mesh with 16 by 16 elements are in perfect agreement with the results presented by Klinkel and colleagues using a similar solid-shell formulation. Our result $u_A = 3.2627$ and $u_B = 5.4763$ for the final displacement at points A and B also agree up to numerical accuracy with the corresponding values reported by Vu-Quoc and Tan [322], $u_A^{[322]} = 3.26055$ and $u_B^{[322]} = 5.48331$, for exactly the same element formulation.

Equally, the results for the mesh with 64 by 64 elements agree with the reference. However, it is clearly observed by the significantly larger displacement that the problem is not converged with the coarser discretization; compare also the results for the 24 by 24 elements. Again the reason lies in the fact that the pinching forces impose a singularity on the problem and the results strongly depend on resolving this singularity.

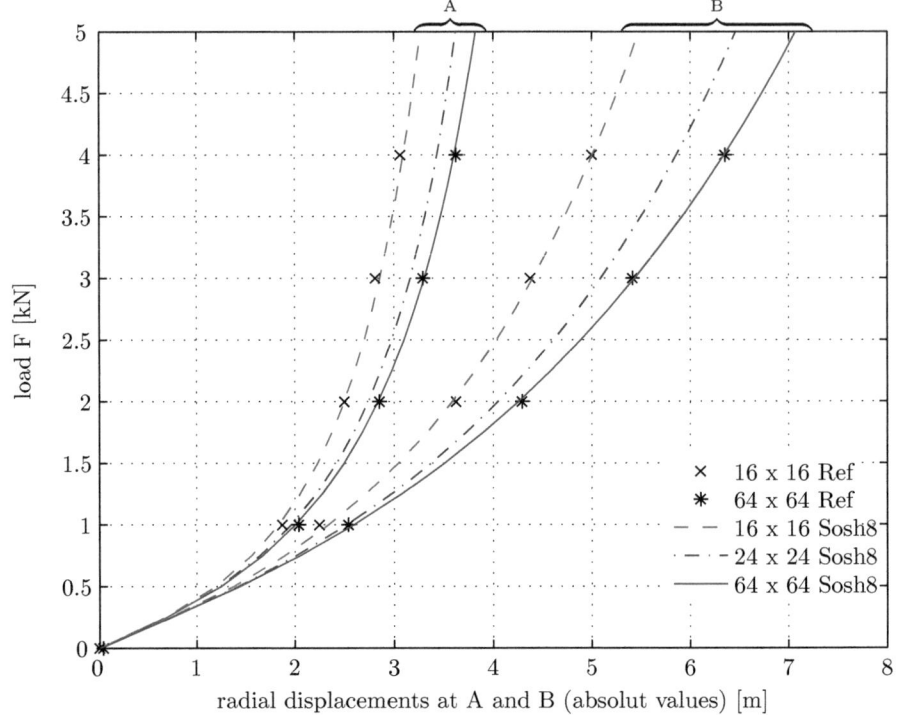

Fig. 3.21: Load-displacement diagram of hexahedral discretizations of the pinched hemisphere.

This issue becomes significant when we look at the results for our proposed Sosh6-element. In the plots of Fig. 3.22 we compare results of different Sosh6-meshes with the results of the Sosh8-discretizations. 'Mesh A' and 'Mesh B' are obtained by subdividing each hexahedron of the 16 by 16 Sosh8-Mesh into two wedges. They differ only in the direction of the diagonal. A third 'Mesh C' results from a free triangular surface mesh with intervals of 16 elements per side, resulting in mostly equilateral triangles and wedges, respectively. The three meshes are depicted in Fig. 3.23 with close-ups for clarification.

The effect of the different mesh cases is intriguing. The results for 'Mesh A' with 512 Sosh6-elements are remarkebly close to the results of the Sosh8-discretization with $64 \times 64 = 4096$ elements. Even for 'Mesh B' they are better than $24 \times 24 = 576$ Sosh8-elements. The reason has to lie in the way, the point-loads are carried into the structure. Apparently, for case 'A' where the node with the inwards pinching load belongs to an edge dividing two wedges is better

3. Efficient Finite Elements for Arterial Wall Modeling

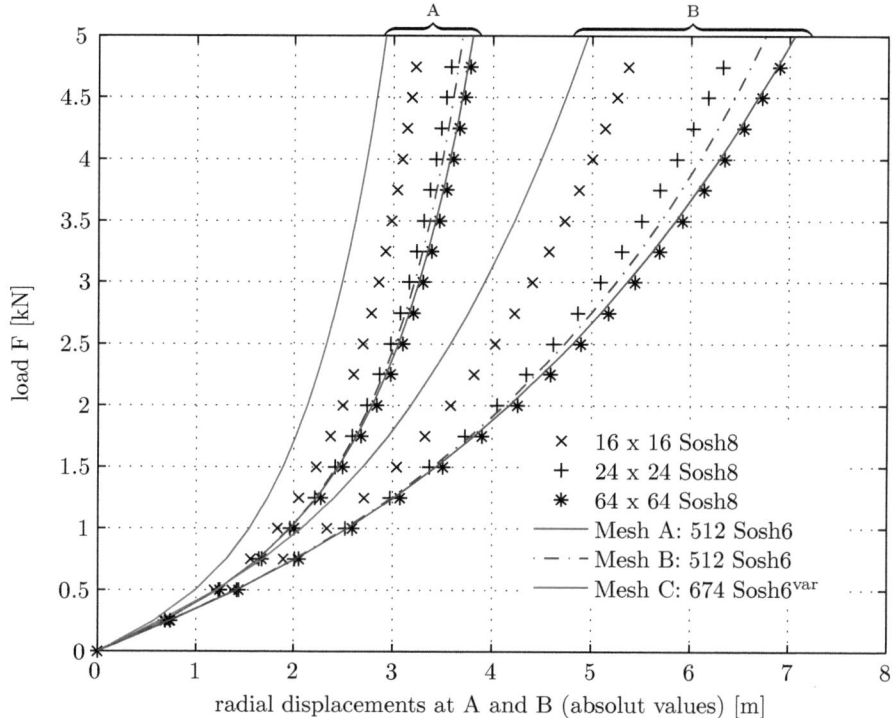

Fig. 3.22: Load-displacement diagram of wedge-shaped discretizations of the pinched hemisphere.

for the Sosh6-discretization with its optimized parameter space. At the node with the outwards pinching load the better element configuration is apparently contrary with the node belonging to just one element. In 'Mesh B' where just the elements are flipped the results are worse. In conclusion, it matters how the pinching forces distribute into the structure and whether the one or the other mesh configuration captures this better or worse. In the close-ups of Fig. 3.23 the slightly different displacements close to the pinching point is noticeable.

Finally, if we obtain the wedge-discretization from a free triangulation of the surface as in 'Mesh C', the optimized parameter space in the DSG formulation does not make sense and is therefore omitted. The corresponding results, Sosh6^{var}, are significantly worse and still influenced by transverse shear locking. Thus, the underlying meshing strategy plays a significant role for the performance of the proposed Sosh6-elements. We foreclose that in our proposed modeling and meshing strategy for patient-specific arterial walls discussed in Chapter 6, we encounter meshes with a majority of rectangular wedges.

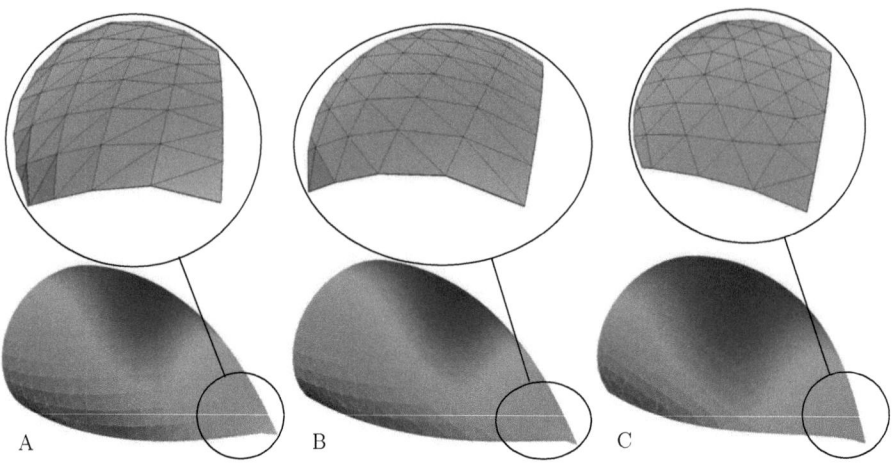

Fig. 3.23: Different wedge meshes for pinched hemisphere with colors representing displacement magnitudes scaled to mesh A.

3.6.4 Cook's Membrane Problem

This final example considers a very famous benchmark for modeling incompressibility where elimination of volumetric locking is essential. Simo and Armero [267] have extended the well-known tapered beam structure, usually used for testing bending dominated problems, to quasi-incompressible finite elasticity. With the applied 'Neo-Hookean' material law, described in more detail in Section 4.2.1, finite elasticity can be modeled correctly. The problem is basically two-dimensional under plain strain conditions, but can be extended into three dimensions using for example a unit thickness. All units are dimensionless as reported in the literature reference. See Fig. 3.24 on the left for a sketch of the structure and the final displacements. As a reference solution we take the results reported by Klinkel *et al.* [171] for a solid-shell element identical to our proposed Sosh8-element. These results are reported to be almost indistinguishable to the original results presented by Simo and Armero for a two-dimensional 'Q1E4' element.

The convergence diagram on the right of Fig. 3.24 prove the success of the EAS method to eliminate volumetric locking. All formulations converge to the same result. The results for the EAS9-elements and the Sosh8-elements are indistinguishable from the reference. The four involved in-plane EAS parameters included in both element formulations resolve the quasi-incompressibility issue. Still, additional EAS parameters included in the EAS21-elements improve convergence.

It was not possible to obtain any results for a discretization applying wedge-shaped elements due to convergence breakup. The problem poses the incompressibility issue solely on the triangular plane of the wedge-elements. As discussed earlier these elements do not apply any element technology within the triangular plane and thus have the same problem with this

3. Efficient Finite Elements for Arterial Wall Modeling

example as pure displacement-based elements. Only if the incompressibility is lowered by reducing Poisson's ratio will produce results, but these are still influenced by volumetric locking.

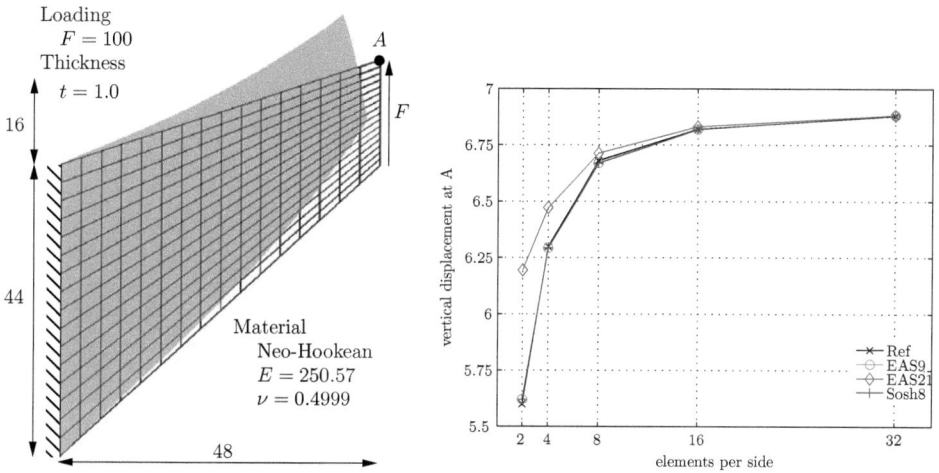

Fig. 3.24: Cook's membrane problem (left) and tip-displacement convergence (right).

3.6.5 Summary

The presented benchmark examples prove the successful implementation of the proposed element techniques resulting in excellent and competitive results. We remark that due to linearization we obtain overall a good convergence behavior with quadratic convergence rates within the convergence radius. But, admittedly, the convergence radius for advanced element techniques is smaller compared to pure displacement-based elements. Also, the matrix condition for three-dimensional solid elements might be worse than for shell elements based on dimensional reduction. In summary, the proposed techniques represent state of the art three-dimensional solid elements and are well-suited to model biomechanical structural problems such as the arterial wall.

4. Constitutive Models for the Arterial Wall

Within this chapter we discuss a selection of constitutive laws which are suited and widely applied for simulations of the arterial wall. The range starts from rather simple formulations which however have a large reputation in biomechanical applications. Subsequently more complex material laws are elaborated which take into account the anisotropic fiber-reinforced microstructure of the wall. All these constitutive models count to the theory of *hyperelasticity*. Further on, an extension to *viscoelasticity* is presented which is typically present in biological tissue and should be considered in a sophisticated study of the arterial wall.

4.1 Essentials of Constitutive Modeling

Prior to the discussion of specified material laws we shortly review the underlying premises of constitutive modeling. In this sense the following section continues the chapter on continuum mechanical foundations (Chapter 2), however having always the application within the finite element method in mind. The topic is extensively elaborated in the literature. Detailed elaborations can be found in the books of Başar [15], Holzapfel [137], Parisch [224], among others.

4.1.1 Properties and Restrictions for Constitutive Laws

For a mechanically and physically consistent constitutive law in terms of an appropriate strain energy function and also a numerically convenient treatment several properties and restrictions have to be fulfilled. We restrict ourselves here to a short overview and refer to the literature for detailed discussions, for instance Truesdell and Noll [312], Marsden and Hughes [208], and Ogden [220]. We first recall some basic principles and then report some beneficial properties for material modeling.

Materials for which the constitutive behavior is only a function of the current state of deformation are generally known as elastic. In the special case when the work done by the stresses during a deformation process is dependent only on the initial state at time t_0 and the final configuration at time t, the behavior of the material is said to be path-independent and the material is termed *hyperelastic*. Such a material postulates the existence of a Helmholtz free-energy function Ψ defined per unit reference volume, see Section 2.3. For the case in which $\Psi = \Psi(F)$ is solely a function of F or some other strain tensor the Helmholtz free-energy function is referred to as the *strain-energy function* or *stored-energy function* or *elastic potential*.

The *principle of determinism and local action* states that the stress of a material point $\mathbf{X}$ occupying $\mathbf{x}$ at the moment t is determined by the past history of the motion of any arbitrarily small neighborhood of $\mathbf{X}$.

The *principle of material objectivity* reveals that the response of a material is the same for all observers. Since this expression is rather cumbersome to evaluate, invariance with respect to superposed rigid body motions is usually required. Thus the strain energy has to be independent to a rigid body rotation, given as

$$\Psi(F) = \Psi(QF) \quad \text{with} \quad Q \in SO(3), \tag{4.1}$$

where $SO(3)$ denotes the special orthogonal group, i.e. $\det Q = 1 \wedge Q^T Q = I$.

The *principle of material symmetry* requires that no rigid body rotation $Q \in \mathcal{G}$ applied to the reference configuration is allowed to alter the material response. Herein, $\mathcal{G}$ identifies the symmetry group of the considered body and characterizes the symmetry properties of the material. If the symmetry group $\mathcal{G}$ is equal to the special orthogonal group $SO(3)$ then the material is called isotropic.

The *concept of internal variables* introduces additional internal material state variables to describe aspects of the internal structure of materials associated with time-dependent behavior, for instance dissipation. The set of internal variables are denoted collectively by $\mathcal{I}$ and their evolution is determined by additional equations. The strain energy then depends in addition to the deformation gradient F on $\mathcal{I}$

$$\Psi = \Psi(F, \mathcal{I}). \tag{4.2}$$

The *properties of the strain energy in special limit cases*, namely the total compression of a continuum body to a single point and the infinite principal strains λ_i of the same body should be consistent with the physical comprehension. This means that in the first case with $\lambda_i \to +0$ the strain energy function satisfies

$$\Psi(\lambda_i \to +0) \to +\infty \tag{4.3}$$

and the principal Cauchy-stresses also tend to infinite pressure values, $\sigma_i \to -\infty$. The second, contrary case $\lambda_i \to +\infty$ requires

$$\Psi(\lambda_i \to +\infty) \to +\infty \tag{4.4}$$

and $\sigma_i \to +\infty$. Additionally it is beneficial, if the reference configuration is stress-free

$$\sigma(F \equiv I) = S(F \equiv I) = 0. \tag{4.5}$$

Polyconvexity in the sense of Ball [16] seems an appropriate condition for hyperelasticity, see also Balzani [17] and references therein. A strain energy function

$$\Psi(\boldsymbol{F}) = P(\boldsymbol{F}, J\boldsymbol{F}^{-T}, J), \qquad (4.6)$$

is defined to be *polyconvex*, if it is convex in its arguments $\boldsymbol{F}, J\boldsymbol{F}^{-T}, J$.

4.1.2 Constitutive Equations in Terms of Invariants

Elastic deformation is defined to be fully reversible. Every state of deformation identified by the deformation gradient $\boldsymbol{F}$ is related with the corresponding strain energy $\Psi(\boldsymbol{F})$. The principle of objectivity (4.1) requires that this strain energy is independent to the observer yielding the relation

$$\Psi = \Psi(\boldsymbol{F}) = \Psi(\boldsymbol{QF}) = \Psi(\boldsymbol{U}) = \Psi(\boldsymbol{C}) = \Psi(\boldsymbol{E}). \qquad (4.7)$$

For isotropic constitutive relations

$$\Psi(\boldsymbol{C}) = \Psi(\boldsymbol{QCQ}) \quad \text{with} \quad \boldsymbol{Q} \in SO(3) \qquad (4.8)$$

must hold. Therefore the strain energy must be independent for arbitrary orthogonal rotations of the right Cauchy-Green tensor. This is ensured, if Ψ depends on the three principal invariants of $\boldsymbol{C}$, which are given as

$$I_1 = \lambda_1^2 + \lambda_2^2 + \lambda_3^2 = \operatorname{tr} \boldsymbol{C} \qquad (4.9)$$

$$I_2 = \lambda_1^2 \lambda_2^2 + \lambda_2^2 \lambda_3^2 + \lambda_3^2 \lambda_1^2 = \frac{1}{2}\left[(\operatorname{tr} \boldsymbol{C})^2 - \operatorname{tr} \boldsymbol{C}^2\right] \qquad (4.10)$$

$$I_3 = \lambda_1^2 \lambda_2^2 \lambda_3^2 = J^2 = \det \boldsymbol{C} \Leftrightarrow J = \sqrt{I_3} \qquad (4.11)$$

where λ_i are the principle strains and λ_i^2 are the eigenvalues of $\boldsymbol{C}$. This results in a formulation of the strain energy in dependence of the invariants

$$\Psi = \Psi(I_1, I_2, I_3) \qquad (4.12)$$

which obeys the requirements of objectivity and allows a convenient and general description of different constitutive laws. In order to satisfy the requirement of a vanishing strain energy in reference configuration the function must obey $\Psi(I_1 = 3, I_2 = 3, I_3 = 1) = 0$. It should be noted that the invariants of the left Cauchy-Green tensor are identical and a formulation $\Psi = \Psi(I_1^b, I_2^b, I_3^b)$ is equivalent.

Stress and elasticity in terms of invariants

The concept of invariants is pursued in the derivation of stress and elasticity tensors. The appropriate tensor derivation of a general strain energy in terms of invariants $\Psi = \Psi(I_1, I_2, I_3)$

yields for the second Piola-Kirchhoff stress

$$\begin{aligned}
\boldsymbol{S} = 2\frac{\partial \Psi(\boldsymbol{C})}{\partial \boldsymbol{C}} &= 2\left(\frac{\partial \Psi}{\partial I_1}\frac{\partial I_1}{\partial \boldsymbol{C}} + \frac{\partial \Psi}{\partial I_2}\frac{\partial I_2}{\partial \boldsymbol{C}} + \frac{\partial \Psi}{\partial I_3}\frac{\partial I_3}{\partial \boldsymbol{C}}\right) \\
&= 2\left(\left(\frac{\partial \Psi}{\partial I_1} + I_1\frac{\partial \Psi}{\partial I_2}\right)\boldsymbol{I} - \frac{\partial \Psi}{\partial I_2}\boldsymbol{C} + I_3\frac{\partial \Psi}{\partial I_3}\boldsymbol{C}^{-1}\right)
\end{aligned} \quad (4.13)$$

Accordingly, the elasticity tensor is obtained with the help of some derivatives with respect to $\boldsymbol{C}$ as

$$\begin{aligned}
\mathbb{C} = 2\frac{\partial \boldsymbol{S}(\boldsymbol{C})}{\partial \boldsymbol{C}} &= 4\frac{\partial^2 \Psi(\boldsymbol{C})}{\partial \boldsymbol{C}\partial \boldsymbol{C}} = 4\frac{\partial^2 \Psi(I_1, I_2, I_3)}{\partial \boldsymbol{C}\partial \boldsymbol{C}} \\
&= 4\Bigg[\left(\frac{\partial^2 \Psi}{\partial I_1 \partial I_1} + 2I_1\frac{\partial^2 \Psi}{\partial I_1 \partial I_2} + \frac{\partial \Psi}{\partial I_2} + I_1^2\frac{\partial^2 \Psi}{\partial I_2 \partial I_2}\right)\boldsymbol{I}\otimes\boldsymbol{I} \\
&\quad - \left(\frac{\partial^2 \Psi}{\partial I_1 \partial I_2} + I_1\frac{\partial^2 \Psi}{\partial I_2 \partial I_2}\right)(\boldsymbol{I}\otimes\boldsymbol{C} + \boldsymbol{C}\otimes\boldsymbol{I}) \\
&\quad + \left(I_3\frac{\partial^2 \Psi}{\partial I_1 \partial I_3} + I_1 I_3\frac{\partial^2 \Psi}{\partial I_2 \partial I_3}\right)(\boldsymbol{I}\otimes\boldsymbol{C}^{-1} + \boldsymbol{C}^{-1}\otimes\boldsymbol{I}) \\
&\quad + \frac{\partial^2 \Psi}{\partial I_2 I_2}\boldsymbol{C}\otimes\boldsymbol{C} - I_3\frac{\partial^2 \Psi}{I_2 I_3}(\boldsymbol{C}\otimes\boldsymbol{C}^{-1} + \boldsymbol{C}^{-1}\otimes\boldsymbol{C}) \\
&\quad + \left(I_3\frac{\partial \Psi}{\partial I_3} + I_3^2\frac{\partial^2 \Psi}{\partial I_3 \partial I_3}\right)\boldsymbol{C}^{-1}\otimes\boldsymbol{C}^{-1} \\
&\quad - I_3\frac{\partial \Psi}{\partial I_3}\boldsymbol{C}^{-1}\boxtimes\boldsymbol{C}^{-1} - \frac{\partial \Psi}{\partial I_2}\boldsymbol{I}\boxtimes\boldsymbol{I}\Bigg].
\end{aligned} \quad (4.14)$$

Herein, the fourth order tensors $-\frac{\partial \boldsymbol{C}^{-1}}{\partial \boldsymbol{C}} = \boldsymbol{C}^{-1}\boxtimes\boldsymbol{C}^{-1}$ and $\boldsymbol{I}\boxtimes\boldsymbol{I}$ are defined according to the special second order tensor product

$$\boldsymbol{A}\boxtimes\boldsymbol{B} = \frac{1}{2}(A_{AC}B_{BD} + A_{AD}B_{BC}). \quad (4.15)$$

4.1.3 Incompressibility and Near Incompressibility

Incompressibility plays an important role in engineering practice. Most large strain processes take place under incompressible or near incompressible conditions. This is true in an elastoplastic context where the plastic deformation is often truly incompressible. Also viscoelasticity is often related to incompressibility. Beyond, numerous polymeric materials sustain large strain without noticeable volume changes. Rubber elasticity was therefore a driving force for research in this field. Biological materials consisting mainly of molecule chains with a high contingent of water usually have to be treated as incompressible. In this section the implications of incompressibility for constitutive equations are described. The consequences for a finite element discretization and associated solution techniques have already been discussed in Chapter 3.

Incompressibility is characterized by the volume constraint

$$J = 1 \quad (4.16)$$

throughout the deformation. Also hydrostatic loading causes no deformation and strain. Therefore, the state of stress cannot be completely derived via a constitutive relation and stress boundary conditions have to be considered. To fully describe the state of stress one additional equation is necessary which is related to the volume constraint, yielding the general relation for the second Piola-Kirchhoff stress:

$$S = 2\frac{\partial \Psi(C)}{\partial C} + pC^{-1}. \tag{4.17}$$

The scalar p serves as an indeterminate Lagrangean multiplier identified as hydrostatic pressure. It may only be determined from equilibrium equations and boundary conditions.

Decoupling of strain energy into isochoric and volumetric parts

The issue of incompressibility does not only pose problems on finite element techniques as discussed in the previous chapter, but has also consequences for an appropriate material description. A general and widely accepted approach to handle this problem is based on a decoupled formulation of the constitutive equations. This means that the volumetric and isochoric response are handled separately via a multiplicative decomposition of the deformation gradient F into volume-changing (dilatational) and volume-preserving (distortional, isochoric) parts:

$$F = F_{\text{vol}} F_{\text{iso}} = (J^{1/3} I) \widehat{F}. \tag{4.18}$$

The benefit is that the two contributions can now be modeled separately where the identification of the isochoric material behavior plays the decisive role in the description of the behavior of the tissue. To model incompressibility any dilatational deformation has to be eliminated. According to the deformation gradient also the right Cauchy-Green strain tensor is decomposed as

$$C = (J^{2/3} I) \widehat{C} = J^{2/3} \widehat{C}. \tag{4.19}$$

In the same way the strain energy can be decoupled into its volumetric part Ψ^{vol} and its isochoric part Ψ^{iso}. The volumetric part depends only on the dilatation J, whereas the isochoric part depends on the modified invariants

$$\Psi = \Psi^{\text{iso}}(\widehat{C}) + \Psi^{\text{vol}}(J) = \Psi^{\text{iso}}(\widehat{I}_1, \widehat{I}_2) + \Psi^{\text{vol}}(J). \tag{4.20}$$

The modified invariants of the right Cauchy-Green tensor are given as

$$\widehat{I}_1(\widehat{C}) = J^{-2/3} I_1 \tag{4.21}$$

$$\widehat{I}_2(\widehat{C}) = J^{-4/3} I_2 \tag{4.22}$$

$$\widehat{I}_3(\widehat{C}) = J^{-2} I_3 \equiv 1. \tag{4.23}$$

Decoupled stress and elasticity

For the derivation of the second Piola-Kirchhoff stress the concept of decoupling into volumetric and isochoric parts leads to an additive split

$$\boldsymbol{S} = 2\frac{\partial \Psi(\boldsymbol{C})}{\partial \boldsymbol{C}} = \boldsymbol{S}^{\text{iso}} + \boldsymbol{S}^{\text{vol}} = 2\frac{\partial \Psi^{\text{iso}}(\widehat{\boldsymbol{C}})}{\partial \boldsymbol{C}} + 2\frac{\partial \Psi^{\text{vol}}(J)}{\partial \boldsymbol{C}}$$

$$= J^{-2/3}\underbrace{\left(\mathbb{I} - \frac{1}{3}\boldsymbol{C}^{-1}\otimes\boldsymbol{C}\right)}_{\mathbb{P}} : \widehat{\boldsymbol{S}} + pJ\boldsymbol{C}^{-1} \qquad (4.24)$$

where we introduced the fourth-order so-called *projection tensor* $\mathbb{P}$ following Holzapfel [137, 136]. Herein the scalar p is the hydrostatic pressure, given as $p = \frac{\text{d}\Psi^{\text{vol}}(J)}{\text{d}J}$ and

$$\widehat{\boldsymbol{S}} = 2\frac{\partial \Psi^{\text{iso}}(\widehat{\boldsymbol{C}})}{\partial \widehat{\boldsymbol{C}}} \qquad (4.25)$$

is the so-called *fictitious second Piola-Kirchhoff stress*. In terms of invariants it is obtained by replacing I_1, I_2 and $\boldsymbol{C}$ with their decoupled isochoric counterparts $\widehat{I}_1, \widehat{I}_2$ and $\widehat{\boldsymbol{C}}$, respectively, and omitting the parts depending on I_3 in (4.13).

Similarly, the elasticity tensor can be formulated in a decoupled additive manner

$$\mathbb{C} = 2\frac{\partial \boldsymbol{S}(\boldsymbol{C})}{\partial \boldsymbol{C}} = \mathbb{C}^{\text{iso}} + \mathbb{C}^{\text{vol}} = 2\frac{\partial \boldsymbol{S}^{\text{iso}}}{\partial \boldsymbol{C}} + 2\frac{\partial \boldsymbol{S}^{\text{vol}}}{\partial \boldsymbol{C}}. \qquad (4.26)$$

The isochoric contribution is, again via a projection proposed by Holzapfel [137, 136], given as

$$\mathbb{C}^{\text{iso}} = J^{-4/3}\,\mathbb{P} : \widehat{\mathbb{C}} : \mathbb{P}^T$$
$$+ \frac{2}{3}J^{-2/3}\left(\widehat{\boldsymbol{S}} : \boldsymbol{C}\right)\left(\boldsymbol{C}^{-1}\boxtimes\boldsymbol{C}^{-1} - \frac{1}{3}\boldsymbol{C}^{-1}\otimes\boldsymbol{C}^{-1}\right)$$
$$- \frac{2}{3}\left(\boldsymbol{C}^{-1}\otimes\boldsymbol{S}^{\text{iso}} + \boldsymbol{S}^{\text{iso}}\otimes\boldsymbol{C}^{-1}\right) \qquad (4.27)$$

with $J^{4/3}\widehat{\mathbb{C}} = 2\frac{\partial \widehat{\boldsymbol{S}}}{\partial \widehat{\boldsymbol{C}}}$ being obtained by replacing I_1, I_2 and $\boldsymbol{C}$ with $\widehat{I}_1, \widehat{I}_2$ and $\widehat{\boldsymbol{C}}$, respectively, and omitting the parts depending on I_3 in (4.14). $\widehat{\mathbb{C}}$ may be called *fictitious elasticity tensor* in the material description. The volumetric contribution yields

$$\mathbb{C}^{\text{vol}} = J\left(p + J\frac{\text{d}p}{\text{d}J}\right)\boldsymbol{C}^{-1}\otimes\boldsymbol{C}^{-1} - 2pJ\boldsymbol{C}^{-1}\boxtimes\boldsymbol{C}^{-1}. \qquad (4.28)$$

Penalty approach to enforce incompressibility

Typically in finite element simulations, a *penalty approach* is employed to enforce incompressibility within a nearly incompressible constitutive law. The dilatational deformation is penalized controlled by the *penalty parameter*. This parameter can usually be identified with the bulk modulus and high values in the range of $10^3 - 10^4\mu$ are common to ensure incompressibility. However, the penalty approach poses numerical problems on the solution strategy which goes along with adaptations of the penalty parameter while monitoring the dilatational deformation.

In an isochore-volumetric decoupled setting the penalty approach only affects the specification of the volumetric strain energy function. There are several proposed functions available in the literature and we refer to the work of Doll [69, 70] and the references therein. A small selection of popular functions is

$$\Psi_1^{\text{vol}}(J) = \frac{\kappa}{2}(J-1)^2, \tag{4.29}$$

$$\Psi_2^{\text{vol}}(J) = \frac{\kappa}{2}(\ln J)^2, \tag{4.30}$$

$$\Psi_3^{\text{vol}}(J) = \frac{\kappa}{2}\left((J-1)^2 + (\ln J)^2\right), \tag{4.31}$$

$$\Psi_4^{\text{vol}}(J) = \frac{\kappa}{4}(J^2 - 1 - 2\ln J). \tag{4.32}$$

Note that not all of these functions satisfy every requirement of Section 4.1.1. For example, Ψ_1^{vol} tends to $\kappa/2$ for $J \to +0$ and should therefore be avoided in applications with large compression, whereas Ψ_2^{vol} does not result in infinite stresses for the case of infinite tension.

Summarizing, the concept of decoupling the material response and its strain energy function together with the penalty approach to enforce incompressibility allows a convenient methodology to model biological tissue in a finite-element solution context. However, several issues, especially in the numerical treatment remain unsolved, as the penalty term is known to yield badly scaled system matrices. Considerations to overcome these problems represent research in progress.

4.1.4 Extension to Anisotropy

The constituents of the arterial wall result in a highly complex material behavior. Since the (wavy) collagen fibers are not active at low pressures (they do not store strain energy) we separate the mechanical response into a non-collagenous matrix material and the response of each collagen fiber family. The resistance of the arterial wall to stretch at high internal pressures is almost entirely due to collagenous fibers. The matrix material is assumed to be isotropic whereas each collagen fiber family introduces a transverse isotropy. This is also a typical setting in other fiber-reinforced materials. It motivates the assumption of an additive composition of the strain energy into an isotropic part for the matrix material and a transversely isotropic part for each fiber family. The general structure of the energy function is then

$$\Psi = \Psi^{\text{blk}} + \Psi^{\text{fib}}. \tag{4.33}$$

With their specific spatial direction these fibers introduce certain anisotropy of the material. This is represented by a particular symmetry group which affects the dependence of the strain energy $\Psi(\mathbf{X}, \mathbf{C})$ of the material. For the case of hyperelastic transversely isotropic materials, the directional dependence of the strain energy function on the deformation is commonly defined by introducing a vector representing the material preferred direction, see Spencer [277]. In order to satisfy the requirements of the *principle of material objectivity* and the *principle of material symmetry* the anisotropic behavior is formulated in a coordinate invariant setting based on the concept of structural tensors. Starting with one specific direction leading to transverse isotropy we introduce one unit vector field **m** describing the local fiber direction of the material in the undeformed configuration.

The strain energy function can then be expressed as an isotropic tensor function of the right Cauchy-Green deformation tensor and the unit vector field **m** as

$$\Psi(\mathbf{X}, \boldsymbol{C}, \mathbf{m}) = \Psi(\mathbf{X}, \boldsymbol{Q}\boldsymbol{C}\boldsymbol{Q}^{\mathrm{T}}, \boldsymbol{Q}\mathbf{m} \otimes \mathbf{m}\boldsymbol{Q}^{\mathrm{T}}) \tag{4.34}$$

for all proper orthogonal tensors $\boldsymbol{Q}$. The corresponding structural tensor $\boldsymbol{M}$ is introduced as

$$\boldsymbol{M} = \mathbf{m} \otimes \mathbf{m} \quad \text{implying} \quad \mathrm{tr}\, \boldsymbol{M} = 1. \tag{4.35}$$

According to Spencer [277] there is an irreducible integrity basis for the symmetric second-order tensors $\boldsymbol{C}$ and $\boldsymbol{M}$, given by the traces of the powers of the argument tensors up to a finite order. We obtain the invariants

$$\begin{aligned} &I_1 = \mathrm{tr}\, \boldsymbol{C}, \quad I_2 = \tfrac{1}{2}\left[(\mathrm{tr}\, \boldsymbol{C})^2 - \mathrm{tr}\, \boldsymbol{C}^2\right], \quad I_3 = \det \boldsymbol{C} \\ &\bar{I}_4 = \mathrm{tr}[\boldsymbol{C}\boldsymbol{M}], \quad \bar{I}_5 = \mathrm{tr}[\boldsymbol{C}^2\boldsymbol{M}], \end{aligned} \tag{4.36}$$

consisting of the standard invariants I_1, I_2, I_3 of $\boldsymbol{C}$ and the non-standard (or pseudo-) invariants $\bar{I}_4$ and $\bar{I}_5$. Because of $\mathrm{tr}\, \boldsymbol{M} = 1$, it holds

$$\bar{I}_{(6)} = \mathrm{tr}[\boldsymbol{C}\boldsymbol{M}^2] = \bar{I}_4 \quad \text{and} \quad \bar{I}_{(7)} = \mathrm{tr}[\boldsymbol{C}^2\boldsymbol{M}^2] = \bar{I}_5 \tag{4.37}$$

providing the construction of transversely isotropic material models with a strain energy function of the form

$$\Psi^{\mathrm{triso}} = \Psi(I_1, I_2, I_3, \bar{I}_4, \bar{I}_5). \tag{4.38}$$

For materials with more than one preferred direction the introduction of further vector fields is straight forward leading to anisotropic material behavior. The arterial wall is typically modeled by layers with helically arranged collagen fibers crossing each other at specific fiber angles. Therefore a second fiber direction is introduced, described in undeformed configuration by the unit vector field **n** and the corresponding structural tensor $\boldsymbol{N} = \mathbf{n} \otimes \mathbf{n}$, yielding

$$\Psi^{\mathrm{aniso}} = \Psi^{\mathrm{blk}}(\boldsymbol{C}) + \Psi^{\mathrm{fib},1}(\boldsymbol{C}, \boldsymbol{M}) + \Psi^{\mathrm{fib},2}(\boldsymbol{C}, \boldsymbol{N}) \tag{4.39}$$

This also results in additional non-standard invariants

$$\bar{I}_6 = \mathrm{tr}[\boldsymbol{C}\boldsymbol{N}], \quad \bar{I}_7 = \mathrm{tr}[\boldsymbol{C}^2\boldsymbol{N}] \tag{4.40}$$

for the second fiber direction and the mixed invariants

$$\bar{I}_8 = \mathbf{m} \cdot \boldsymbol{C}\mathbf{n} \quad \text{and} \quad \bar{I}_9 = (\mathbf{m} \cdot \mathbf{n})^2. \tag{4.41}$$

It is worth mentioning that the invariant $\bar{I}_4$ can be reformulated as

$$\bar{I}_4 = \mathbf{m} \cdot \boldsymbol{C}\mathbf{m} = \boldsymbol{F}\mathbf{m} \cdot \boldsymbol{F}\mathbf{m} = \lambda_m^2 \tag{4.42}$$

4. Constitutive Models for the Arterial Wall

with $\lambda_m \tilde{\mathbf{m}} = \mathbf{F}\mathbf{m}$ as the map of the undeformed unit vector $\mathbf{m}$ into the current configuration $\tilde{\mathbf{m}}$ and thus as the stretch of the corresponding fiber. It has therefore a clear physical meaning. The same holds for the invariant $\bar{I}_6$ representing the stretch in the second fiber direction. However, the invariants $\bar{I}_5$, $\bar{I}_7$ and $\bar{I}_8$ lack such a physical interpretation and also a high correlation to each other exists which makes experimental identification difficult (see Peña et al. [227]). Further, $\bar{I}_9$ does not depend on the deformation but just on the geometrical layout of the two fiber directions. They are therefore omitted in many constitutive models for arterial walls and a strain energy function

$$\Psi^{\text{aniso}} = \Psi(I_1, I_2, I_3, \bar{I}_4, \bar{I}_6). \tag{4.43}$$

is employed.

Decoupling into isochoric and volumetric deformation

To specifically account for the incompressibility constraint the multiplicative decomposition of the deformation gradient (4.18) is applied. This leads to the *decoupled* representation of the strain energy function of a fiber-reinforced anisotropic material

$$\begin{aligned}\Psi^{\text{aniso}} &= \Psi^{\text{vol}}(J) + \Psi^{\text{aniso}}(\widehat{\mathbf{C}}, \mathbf{M}, \mathbf{N}) \\ &= \Psi^{\text{vol}}(J) + \Psi^{\text{aniso}}(\widehat{I}_{1,2}(\widehat{\mathbf{C}}), \widehat{I}_{4,5}(\mathbf{M}, \widehat{\mathbf{C}}), \widehat{I}_{6,7}(\mathbf{N}, \widehat{\mathbf{C}}))\end{aligned} \tag{4.44}$$

where the modified invariants $\widehat{I}_1, \widehat{I}_2$ are derived according to section 4.1.3 and the remaining modified non-standard invariants are expressed by

$$\widehat{I}_4 = J^{-2/3}\bar{I}_4, \qquad\qquad \widehat{I}_5 = J^{-4/3}\bar{I}_5 \tag{4.45}$$

$$\widehat{I}_6 = J^{-2/3}\bar{I}_6, \qquad\qquad \widehat{I}_7 = J^{-4/3}\bar{I}_7. \tag{4.46}$$

To combine both previous decoupling strategies and thus further modularize the strain energy function for the application in arterial wall models we additively split the isochoric part into an isotropic part for describing the ground substance and an anisotropic part describing each fiber family.

$$\Psi^{\text{aniso}} = \Psi^{\text{vol}}(J) + \Psi^{\text{blk}}(\widehat{\mathbf{C}}) + \Psi^{\text{fib,1}}(\widehat{\mathbf{C}}, \mathbf{M}) + \Psi^{\text{fib,2}}(\widehat{\mathbf{C}}, \mathbf{N}) \tag{4.47}$$

Stress and elasticity tensor

The general derivation of stress and elasticity for isochore-volumetric decoupled material formulations in terms of invariants is presented earlier in this chapter. We extend these equations here for anisotropy, still in the decoupled approach and in terms of the decoupled non-standard invariants

$$\Psi^{\text{aniso}} = \Psi^{\text{vol}}(J) + \Psi^{\text{blk}}(\widehat{I}_1, \widehat{I}_2) + \Psi^{\text{fib,1}}(\widehat{I}_1, \widehat{I}_2, \widehat{I}_4, \widehat{I}_5) + \Psi^{\text{fib,2}}(\widehat{I}_1, \widehat{I}_2, \widehat{I}_6, \widehat{I}_7). \tag{4.48}$$

Applying the chain rule the general expression for the second Piola-Kirchhoff stress tensor is obtained

$$\boldsymbol{S} = 2 \sum_k \left(\frac{\partial \Psi}{\partial L_k} \frac{\partial L_k}{\partial \boldsymbol{C}} \right) \quad \text{with} \quad L_k \in \mathcal{P}, \tag{4.49}$$

where $\mathcal{P}$ is the corresponding polynomial basis. In the modularized decoupled representation of the strain energy both stress and elasticity tensors retain the same additive decomposition yielding for the second Piola-Kirchhoff stress

$$\begin{aligned} \boldsymbol{S} &= \boldsymbol{S}^{\text{vol}} + \boldsymbol{S}^{\text{blk}} + \boldsymbol{S}^{\text{fib},1} + \boldsymbol{S}^{\text{fib},2} \\ &= 2 \left(\frac{\partial \Psi^{\text{vol}}(J)}{\partial \boldsymbol{C}} + \frac{\partial \Psi^{\text{blk}}(\widehat{\boldsymbol{C}})}{\partial \boldsymbol{C}} + \frac{\partial \Psi^{\text{fib},1}(\widehat{\boldsymbol{C}}, \boldsymbol{M})}{\partial \boldsymbol{C}} + \frac{\partial \Psi^{\text{fib},2}(\widehat{\boldsymbol{C}}, \boldsymbol{N})}{\partial \boldsymbol{C}} \right). \end{aligned} \tag{4.50}$$

We refer to (4.24) for the derivation of the isotropic parts and present here just the anisotropic part. Therefore, we add to the *fictitious second Piola-Kirchhoff stress* the anisotropic part $\widehat{\boldsymbol{S}} = \widehat{\boldsymbol{S}}^{\text{iso}} + \widehat{\boldsymbol{S}}^{\text{fib},1} + \widehat{\boldsymbol{S}}^{\text{fib},2}$ where

$$\widehat{\boldsymbol{S}}^{\text{fib},1} = 2 \left[\left(\frac{\partial \Psi^{\text{fib},1}}{\partial \widehat{I}_1} + \widehat{I}_1 \frac{\partial \Psi^{\text{fib},1}}{\partial \widehat{I}_2} \right) \boldsymbol{I} - \frac{\partial \Psi^{\text{fib},1}}{\partial \widehat{I}_2} \widehat{\boldsymbol{C}} + \frac{\partial \Psi^{\text{fib},1}}{\partial \widehat{I}_4} \boldsymbol{M} + \frac{\partial \Psi^{\text{fib},1}}{\partial \widehat{I}_5} \widehat{\boldsymbol{CM}} \right] \tag{4.51}$$

with the second-order tensor $\widehat{\boldsymbol{CM}} = \widehat{\boldsymbol{C}}\boldsymbol{M} + \boldsymbol{M}\widehat{\boldsymbol{C}}$ and $\widehat{\boldsymbol{S}}^{\text{fib},2}$ to be obtained equivalently.

Similarly, the elasticity tensor is obtained by adding the anisotropic contribution to the *fictitious elasticity*, replacing $\widehat{\mathbb{C}}$ in (4.27) with

$$\widehat{\mathbb{C}} = \widehat{\mathbb{C}}^{\text{iso}} + \widehat{\mathbb{C}}^{\text{fib},1} + \widehat{\mathbb{C}}^{\text{fib},2} \tag{4.52}$$

where

$$\begin{aligned} \widehat{\mathbb{C}}^{\text{fib},1} = 4 \bigg[& \left(\frac{\partial^2 \Psi^{\text{fib},1}}{\partial \widehat{I}_1 \partial \widehat{I}_4} + \widehat{I}_1 \frac{\partial^2 \Psi^{\text{fib},1}}{\partial \widehat{I}_2 \partial \widehat{I}_4} \right) (\boldsymbol{I} \otimes \boldsymbol{M} + \boldsymbol{M} \otimes \boldsymbol{I}) \\ & - \frac{\partial^2 \Psi^{\text{fib},1}}{\partial \widehat{I}_2 \partial \widehat{I}_4} \left(\widehat{\boldsymbol{C}} \otimes \boldsymbol{M} + \boldsymbol{M} \otimes \widehat{\boldsymbol{C}} \right) + \frac{\partial^2 \Psi^{\text{fib},1}}{\partial \widehat{I}_4 \partial \widehat{I}_4} \boldsymbol{M} \otimes \boldsymbol{M} \\ & + \left(\frac{\partial^2 \Psi^{\text{fib},1}}{\partial \widehat{I}_1 \partial \widehat{I}_5} + \widehat{I}_1 \frac{\partial^2 \Psi^{\text{fib},1}}{\partial \widehat{I}_2 \partial \widehat{I}_5} \right) \left(\boldsymbol{I} \otimes \widehat{\boldsymbol{CM}} + \widehat{\boldsymbol{CM}} \otimes \boldsymbol{I} \right) \\ & - \frac{\partial^2 \Psi^{\text{fib},1}}{\partial \widehat{I}_2 \partial \widehat{I}_5} \left(\widehat{\boldsymbol{C}} \otimes \widehat{\boldsymbol{CM}} + \widehat{\boldsymbol{CM}} \otimes \widehat{\boldsymbol{C}} \right) + \frac{\partial^2 \Psi^{\text{fib},1}}{\partial \widehat{I}_5 \partial \widehat{I}_5} \left(\widehat{\boldsymbol{CM}} \otimes \widehat{\boldsymbol{CM}} \right) \\ & + \frac{\partial^2 \Psi^{\text{fib},1}}{\partial \widehat{I}_4 \partial \widehat{I}_5} \left(\boldsymbol{M} \otimes \widehat{\boldsymbol{CM}} + \widehat{\boldsymbol{CM}} \otimes \boldsymbol{M} \right) \\ & + \frac{\partial \Psi^{\text{fib},1}}{\partial \widehat{I}_5} \{ \boldsymbol{I} \boxtimes \boldsymbol{M} + \boldsymbol{M} \boxtimes \boldsymbol{I} \} \bigg] \end{aligned} \tag{4.53}$$

and again, $\widehat{\mathbb{C}}^{\text{fib},2}$ is obtained equivalently.

4.2 Basic Isotropic Models

Two rather simple hyperelastic material laws, a *Neo-Hookean*-type material and a *Mooney-Rivlin*-type material are introduced within this section. Both are suited to model large strain deformations and can often be found in the field of biomechanical simulations.

4.2.1 Neo-Hookean Material

The Neo-Hookean-type material is one of the simplest hyperelastic constitutive laws. It is suited for large deformations and strains and recovers linear elasticity in the small deformation case. It is based on only two parameters, for instance Young's modulus E and Poisson's ratio ν, or shear modulus μ and bulk modulus κ. Therefore, experimental parameter identification is notably simple and it is often found in complex biomechanical problems like fluid-structure-interaction simulations where the focus does not lie on the material description. Additionally, it usually serves in complex anisotropic material laws to model the isotropic ground substance of the tissue (see later sections).

For instance a form of a *coupled* Neo-Hookean-type strain energy in terms of the Lamé parameters μ and Λ is

$$\Psi(\boldsymbol{C}) = \frac{\mu}{2}(I_1 - 3) - \mu \ln J + \frac{\Lambda}{2}(\ln J)^2 \quad (4.54)$$

where the coupling becomes obvious in the mixed term $\mu \ln J$ and the dependency of the invariant I_1 of the complete strain tensor $\boldsymbol{C}$.

A decoupled representation of a Neo-Hookean-type strain energy function with the quadratic function Ψ_1^{vol} (4.29) for the volumetric part reads

$$\Psi^{\text{NH}} = \frac{\mu}{2}(\widehat{I}_1 - 3) + \frac{\kappa}{2}(J - 1)^2 \quad (4.55)$$

According to (4.24) we find the second Piola-Kirchhoff stress tensor for the Neo-Hookean material (4.55)

$$\boldsymbol{S}^{\text{NH}} = \mu J^{-2/3}\left(\boldsymbol{I} - \frac{1}{3}I_1 \boldsymbol{C}^{-1}\right) + \kappa(J - 1)J\boldsymbol{C}^{-1}. \quad (4.56)$$

The material elasticity tensor reads

$$\mathbb{C}^{\text{NH}} = 2\mu J^{-2/3}\left(\tfrac{1}{3}I_1 \boldsymbol{C}^{-1} \boxtimes \boldsymbol{C}^{-1} - \tfrac{1}{3}\boldsymbol{I} \otimes \boldsymbol{C}^{-1} - \tfrac{1}{3}\boldsymbol{C}^{-1} \otimes \boldsymbol{I} + \tfrac{1}{9}I_1 \boldsymbol{C}^{-1} \otimes \boldsymbol{C}^{-1}\right)$$
$$+ \kappa(J^2 - J)\left(\boldsymbol{C}^{-1} \otimes \boldsymbol{C}^{-1} - 2\boldsymbol{C}^{-1} \boxtimes \boldsymbol{C}^{-1}\right) + \kappa J^2 \boldsymbol{C}^{-1} \otimes \boldsymbol{C}^{-1} \quad (4.57)$$

4.2.2 Mooney-Rivlin Material Law

Another prevalent isotropic material law suited for large strains is the Mooney-Rivlin material law. It is quite successful in modeling elastomeric materials such as rubber, but is also popular in biomechanical applications. Compared to the Neo-Hooke material the Mooney-Rivlin material is characterized by stronger stiffening for large strains, especially in biaxial loading conditions.

Thus, the stiffening of elastomeric materials due to the stretch of the molecule chains can be better approximated.

The general form of strain energy functions assuming incompressibility attributable to Mooney and Rivlin is expressed as

$$\Psi(\boldsymbol{C}) = \sum_{r,s \geq 0} \mu_{rs}(I_1 - 3)^r (I_2 - 3)^s. \tag{4.58}$$

A very frequently employed member of this family is represented by

$$\begin{aligned}\Psi(\boldsymbol{C}) &= \frac{\mu_1}{2}(I_1 - 3) + \frac{\mu_2}{2}(I_2 - 3) \\ &= \frac{\mu_1}{2}(\lambda_1^2 + \lambda_2^2 + \lambda_3^2 - 3) + \frac{\mu_2}{2}(\lambda_1^{-2} + \lambda_2^{-2} + \lambda_3^{-2} - 3)\end{aligned} \tag{4.59}$$

To model nearly incompressible behavior we present another member, referring to Klinkel et al. [172],

$$\Psi^{\text{MR, Klinkel}} = \frac{\mu_1}{2}(I_1 - 3) + \frac{\mu_2}{2}(I_2 - 3) - (\mu_1 + \mu_2)\ln J + \frac{\Lambda}{4}(J^2 - 1 - 2\ln J) \tag{4.60}$$

with Λ being the Lamé constant which may be obtained from the shear modulus $\mu = \mu_1 + \mu_2$ and the bulk modulus κ as $\Lambda = \kappa - 2/3\mu$. It serves here as penalty parameter to enforce incompressibility. It should be remarked that this formulation is not stress-free in reference configuration. Therefore, we slightly modified this function obtaining a stress-free reference configuration

$$\Psi^{\text{MR}} = \frac{\mu_1}{2}(I_1 - 3) + \frac{\mu_2}{2}(I_2 - 3) - (\mu_1 + 2\mu_2)\ln J + \kappa(J - 1)^2. \tag{4.61}$$

The corresponding second Piola-Kirchhoff stress is given as

$$\boldsymbol{S}^{\text{MR}} = (\mu_1 + \mu_2 I_1)\boldsymbol{I} - \mu_2 \boldsymbol{C} + (2\kappa J(J-1) - (\mu_1 + 2\mu_2))\boldsymbol{C}^{-1} \tag{4.62}$$

and the elasticity yields

$$\begin{aligned}\mathbb{C}^{\text{MR}} &= 2\mu_1 \boldsymbol{I} \otimes \boldsymbol{I} - 2\mu_1 \boldsymbol{I} \boxtimes \boldsymbol{I} + 2\kappa J(2J-1)\boldsymbol{C}^{-1} \otimes \boldsymbol{C}^{-1} \\ &\quad - (4\kappa J(J-1) - 2(\mu_1 + 2\mu_2))\boldsymbol{C}^{-1} \boxtimes \boldsymbol{C}^{-1}.\end{aligned} \tag{4.63}$$

A decoupled Mooney-Rivlin-type formulation would be straightforward.

4.3 A Family of Anisotropic Models

4.3.1 Holzapfel's Model

The group of Holzapfel has contributed to the research of the structural behavior of arterial walls in quite a significant way. They propose a constitutive framework adopted by many researchers and the notion 'Holzapfel'-model is commonly applied in the community. The initial idea presented by Holzapfel and Weizsäcker [146] is to extend the famous 'Fung'-model

4. Constitutive Models for the Arterial Wall

proposed by Fung et al. [106] to capture the S-shaped stress-strain behavior of arteries in circumferential direction. Therefore, they add a Neo-Hookean part to the exponential function of the Fung-model and thus allow to separately model the isotropic response of the elastin and ground substance and the anisotropic character of the collagen fibers, reflected in

$$\Psi = \Psi^{\text{iso}} + \Psi^{\text{aniso}}. \tag{4.64}$$

In Holzapfel, Gasser and Ogden [140] they present a profound constitutive framework for arterial walls and also elaborate their proposed anisotropic model. Due to the sound continuum-mechanical basis in terms of invariants it satisfies all major requirements for constitutive laws. It has been extended in several ways, see the following sections. In this section the original model is discussed and the necessary stress and elasticity tensors for the finite-element implementation are reviewed.

Taking into account the incompressibility of the arterial wall tissue we follow the approach of decoupling into volumetric and isochoric deformation. As mentioned the ground substance is modeled by a Neo-Hookean material (see Section 4.2.1), yielding a contribution

$$\Psi^{\text{blk}} = \frac{c}{2}\left(\widehat{I}_1 - 3\right) \tag{4.65}$$

with the stress-like material parameter c.

The strong stiffening effect due to recruitment of collagen observed at high pressures motivated the use of an exponential function for the description of the strain energy stored in the collagen fibers. Taking into account two helically arranged fiber families characterized by the two structural tensors $\boldsymbol{M}$ and $\boldsymbol{N}$ each with equal structural behavior they propose

$$\Psi^{\text{fib1,2}} = \frac{k_1}{2k_2} \sum_{i=4,6} \left(\exp\left(k_2(\widehat{I}_i - 1)^2\right) - 1\right) \tag{4.66}$$

for the anisotropic part, with $k_1 > 0$ as stress-like parameter and $k_2 > 0$ as dimensionless parameter and $\widehat{I}_{4,6}$ as non-standard invariants (see (4.45) and (4.46) of Section 4.1.4). It is assumed that the collagen fibers contribute only in tension, therefore the corresponding terms are omitted if $\widehat{I}_4 \leq 0$ and $\widehat{I}_6 \leq 0$, respectively.

The complete strain-energy then reads

$$\begin{aligned}\Psi^{\text{HGO}} &= \Psi^{\text{blk}}(\widehat{I}_1) + \Psi^{\text{fib,1}}(\widehat{I}_4) + \Psi^{\text{fib,2}}(\widehat{I}_6) + \Psi^{\text{vol}}(J) \\ &= \frac{c}{2}\left(\widehat{I}_1 - 3\right) + \frac{k_1}{2k_2}\sum_{i=4,6}\left(\exp\left(k_2(\widehat{I}_i - 1)^2\right) - 1\right) + \frac{\kappa}{2}(J-1)^2\end{aligned} \tag{4.67}$$

where we specified the volumetric part with Ψ_1^{vol} (4.29) as penalty function to enforce incompressibility.

Stress and elasticity reflect the same additive composition

$$\boldsymbol{S}^{\text{HGO}} = \boldsymbol{S}^{\text{blk}} + \boldsymbol{S}^{\text{fib,1}} + \boldsymbol{S}^{\text{fib,2}} + \boldsymbol{S}^{\text{vol}} \tag{4.68}$$

$$\mathbb{C}^{\text{HGO}} = \mathbb{C}^{\text{blk}} + \mathbb{C}^{\text{fib,1}} + \mathbb{C}^{\text{fib,2}} + \mathbb{C}^{\text{vol}} \tag{4.69}$$

and we particularize $\widehat{\boldsymbol{S}}^{\text{fib},1}$ in (4.51) for convenience as

$$\widehat{\boldsymbol{S}}^{\text{fib},1}_{\text{HGO}} = 2k_1(\widehat{I}_4 - 1)\exp\left(k_2(\widehat{I}_4 - 1)^2\right)\boldsymbol{M} \quad \text{for} \quad \widehat{I}_4 > 0, \tag{4.70}$$

as well as $\widehat{\mathbb{C}}^{\text{fib},1}$ in (4.53) as

$$\widehat{\mathbb{C}}^{\text{fib},1}_{\text{HGO}} = 4\left(k_1 + 2k_1k_2(\widehat{I}_4 - 1)^2\right)\exp\left(k_2(\widehat{I}_4 - 1)^2\right)\boldsymbol{M} \otimes \boldsymbol{M}. \tag{4.71}$$

4.3.2 Balzani's Model

The concept of polyconvexity, introduced by Ball [16] plays an important role in solving the underlying boundary value problems of finite elasticity. The group of Schröder and colleagues has significantly contributed to the research in this field and a variety of polyconvex functions for transversely isotropic and orthotropic materials is proposed in Schröder and Neff [258, 259], see also Schröder et al. [260]. A view to material stability and the adjustment to experimental data for biological tissue is presented by Balzani et al. [19]. The emphasis on polyconvexity is also seized by Itskov et al. [160] who propose a strain-energy function composed of a series of functions and associated parameters sets, each satisfying the polyconvexity conditions.

A particular strain-energy function suited for arterial wall tissue and with proven polyconvexity proposed by Balzani [17] is discussed in the following. The isotropic matrix material is given by

$$\Psi^{\text{iso}} = c_1\left(\widehat{I}_1 - 3\right) + \epsilon_1\left(I_3^{\epsilon_2} + \frac{1}{I_3^{\epsilon_2}} - 2\right), \quad c_1 > 0, \ \epsilon_1 > 0, \ \epsilon_2 > 1, \tag{4.72}$$

where the first part is a decoupled Neo-Hookean type function with material parameter c_1 and the second part is a penalty term to control the volumetric deformation with the penalty parameters ϵ_1 and ϵ_2.

The strain energy for one fiber family is assumed as

$$\Psi^{\text{fib},1} = \begin{cases} \alpha_1(K_1 - 2)^{\alpha_2} & \text{for } K_1 > 2 \\ 0 & \text{for } K_1 \leq 2, \end{cases} \tag{4.73}$$

with the second case excluding pressure in the fibers. For the description of the fiber material another invariant is introduced as

$$K_1 := \text{tr}[\text{cof}[\boldsymbol{C}](\boldsymbol{I} - \boldsymbol{M})] = I_1\bar{I}_4 - \bar{I}_5 \tag{4.74}$$

where $\text{cof}[\boldsymbol{C}] = \det(\boldsymbol{C})\boldsymbol{C}^{-1}$ and which also ensures polyconvexity of the strain-energy. The parameter $\alpha_1 > 0$ defines the initial slope of the fiber stiffness and the parameters $\alpha_2 > 2$ controls the stiffening of the fibers for larger strains. A second fiber family may be introduced accordingly with $K_2 := I_1\bar{I}_6 - \bar{I}_7$ as

$$\Psi^{\text{fib},2} = \begin{cases} \alpha_3(K_2 - 2)^{\alpha_4} & \text{for } K_2 > 2 \\ 0 & \text{for } K_2 \leq 2, \end{cases} \tag{4.75}$$

4. Constitutive Models for the Arterial Wall

again with $\alpha_3 > 0$ and $\alpha_4 > 2$. The full strain-energy of the anisotropic fiber-reinforced material law reads then

$$\Psi^{\text{Balz}} = c_1 \left(\widehat{I}_1 - 3\right) + \epsilon_1 \left(I_3^{\epsilon_2} + \frac{1}{I_3^{\epsilon_2}} - 2\right) + \alpha_1 (K_1 - 2)^{\alpha_2} + \alpha_3 (K_2 - 2)^{\alpha_4} \quad (4.76)$$

The second Piola-Kirchhoff stress is obtained as

$$\begin{aligned} \boldsymbol{S}^{\text{Balz}} = & \frac{1}{I_3} \left(-\tfrac{1}{3} c_1 \widehat{I}_1 + \epsilon_1 \epsilon_2 \left(I_3^{\epsilon_2} - I_3^{-\epsilon_2}\right)\right) \boldsymbol{I} \\ & + \alpha_1 \alpha_2 (K_3 - 2)^{(\alpha_2 - 1)} \left(I_1 \boldsymbol{M} - \boldsymbol{C}^{\boldsymbol{M}} + \bar{I}_4 \boldsymbol{I}\right) \end{aligned} \quad (4.77)$$

where we omitted the second fiber family for the sake of clarity. The second-order tensor $\boldsymbol{C}^{\boldsymbol{M}}$ is defined as $\boldsymbol{C}^{\boldsymbol{M}} = \boldsymbol{C}\boldsymbol{M} + \boldsymbol{M}\boldsymbol{C}$. The corresponding elasticity tensor again omitting the second fiber family reads

$$\begin{aligned} \mathbb{C}^{\text{Balz}} = & \left(\tfrac{4}{9} c_1 \widehat{I}_1 + \epsilon_1 \epsilon_2 \left((\epsilon_2 - 1) I_3^{\epsilon_2} + (\epsilon_2 + 1) I_3^{-\epsilon_2}\right)\right) \boldsymbol{C}^{-1} \otimes \boldsymbol{C}^{-1} \\ & - \left(\tfrac{1}{3} c_1 \widehat{I}_1\right) \left(\boldsymbol{I} \otimes \boldsymbol{C}^{-1} + \boldsymbol{C}^{-1} \otimes \boldsymbol{I}\right) \\ & - \left(c_1 \widehat{I}_1 + \epsilon_1 \epsilon_2 \left(I_3^{\epsilon_2} - I_3^{-\epsilon_2}\right)\right) \left(\boldsymbol{C}^{-1} \otimes \boldsymbol{C}^{-1} + \boldsymbol{C}^{-1} \boxtimes \boldsymbol{C}^{-1}\right) \\ & + \left[\bar{I}_4^2 (\alpha_2 - 1) \boldsymbol{I} \otimes \boldsymbol{I} + I_1^2 (\alpha_2 - 1) \boldsymbol{M} \otimes \boldsymbol{M} + (\alpha_2 - 1) \boldsymbol{C}^{\boldsymbol{M}} \otimes \boldsymbol{C}^{\boldsymbol{M}}\right. \\ & + \left(I_1 \bar{I}_4 (\alpha_2 - 1) + K_3 - 2\right) \left(\boldsymbol{I} \otimes \boldsymbol{M} + \boldsymbol{M} \otimes \boldsymbol{I}\right) \\ & + \bar{I}_4 (1 - \alpha_2) \left(\boldsymbol{C}^{\boldsymbol{M}} \otimes \boldsymbol{I} + \boldsymbol{I} \otimes \boldsymbol{C}^{\boldsymbol{M}}\right) \\ & + I_1 (1 - \alpha_2) \left(\boldsymbol{C}^{\boldsymbol{M}} \otimes \boldsymbol{M} + \boldsymbol{M} \otimes \boldsymbol{C}^{\boldsymbol{M}}\right) \\ & \left. - (K_3 - 2) \{\boldsymbol{I} \boxtimes \boldsymbol{M} + \boldsymbol{M} \boxtimes \boldsymbol{I}\}\right] \alpha_1 \alpha_2 (K_3 - 2)^{(\alpha_2 - 2)}. \end{aligned} \quad (4.78)$$

It is remarked that for one fiber family the invariant K_1 depends on the isotropic invariant I_1 and both modified invariants $\bar{I}_4, \bar{I}_5$ which are not decoupled from the volumetric deformation. Also the specific influence of $\bar{I}_4$ and $\bar{I}_5$ lacks a physical interpretation. This probably complicates parameter identification in experimental setups in comparison to the fully modularized framework favored in Section 4.1.4.

4.3.3 Modified Anisotropic Models

Fiber dispersion is taken into account by Gasser et al. [110] as they modify the Holzapfel-model. Under fiber dispersion a large deviation of individual collagen fibers from the mean orientation of the two helically arranged fiber families is understood. Using polarized light microscopy the group of Canham, Finlay and colleagues [52],[89] have found such a dispersed fiber arrangement in the adventitial layers of small arteries like brain and coronary arteries.

To account of such a structural behavior Gasser incorporate one additional scalar parameter κ_{disp} into the strain-energy ψ^{HGO} (4.67) yielding for the fiber part

$$\Psi^{\text{fib,disp}} = \frac{k_1}{2k_2} \sum_{i=4,6} \left(\exp\left(k_2 (\kappa_{\text{disp}} \widehat{I}_1 + (1 - 3\kappa_{\text{disp}}) \widehat{I}_i - 1)^2\right) - 1\right) \quad (4.79)$$

with $\kappa_{\text{disp}} \in [0; \frac{1}{3}]$. The limit case $\kappa_{\text{disp}} = 0$ is equivalent with the original strain-energy function ψ^{HGO} (4.67) where all fibers of one family are perfectly aligned with directions specified in $\boldsymbol{M}$ and $\boldsymbol{N}$.

In contrast, the other limit case $\kappa_{\text{disp}} = \frac{1}{3}$ corresponds to a completely isotropic distribution of the fibers. A similar exponential-type isotropic strain-energy function has been proposed earlier by Demiray [65] and Delfino et al. [64] who by the way present one of the first sophisticated finite elements analysis of the carotid bifurcation.

Consideration of fiber recruitment is an important question and therefore another model conceptually similar to the Holzapfel-model is presented by Zulliger et al. [343]. Early histological investigation of arterial tissue has shown that collagen fibers are crimped in the unstressed state (see for instance Roach and Burton [248] and Lanir [190]). In their model Zulliger and colleagues incorporate the waviness of the collagen fibers by a statistical distribution function (see also Wuyts et al. [337]). Therein a specified 'engagement strain' based on a log-logistic distribution controls the unfolding of collagen fibers. However, the improvement of a material parameter fit presented in their publication is small and the numerical effort for such a constitutive law seems unprofitable.

In a recent publication by Rodriguez et al. [251] both issues of fiber dispersion and fiber crimping are considered in a modified Holzapfel model. A strain-energy function for one fiber family of

$$\Psi^{\text{fib},1} = \frac{k_1}{2k_2} \left(\exp\left(k_2((1-\rho)(\widehat{I}_1 - 3)^2 + \rho(\widehat{I}_4 - \widehat{I}_4^0)^2\right) - 1 \right) \quad (4.80)$$

is applied where the weighting parameter $\rho \in [0; 1]$ controls the amount of anisotropy and thus the fiber dispersion. The parameter $\widehat{I}_4^0 > 1$ is regarded as the initial crimping of the fibers and the corresponding anisotropic term only contributes for $\widehat{I}_4 > \widehat{I}_4^0$. However, two additional material properties only for one fiber family contribution are introduced doubling the initial number and experimental fitting becomes significantly more difficult.

Multiple layers of collagenous tissue are considered in a recent modification of the Holzapfel-model, published by Kroon and Holzapfel [181]. It is known that many biomechanical tissues consist of a layered structure of collagen fibers. The arterial wall structure is described in detail in Section 1.1 but we repeat here that even the three main layers, Intima, Media, and Adventitia themselves consist of several (up to 70) fenestrated elastic laminae. In their model Kroon and Holzapfel take this layered structure into account and allow each layer of collagen fibers to correspond to a specific fiber direction and a specific fiber stiffness. Therefore, the fiber part of the strain energy consists of a sum of each fiber family contribution where one family itself is modeled as before with an exponential function but pre-multiplied by the corresponding stiffness parameter and evaluated with the corresponding alignment angle. The constitutive equation based on this layered structure is then included into a membrane model of the arterial wall and in their paper the material is fitted to model adventitial extension-inflation behavior using eight layers.

For our proposed arterial wall model we adopt the idea of taking several layers into account. However, in contrast to the model proposed by Kroon and Holzapfel we prefer to model each layer structurally with a corresponding finite element layer. This concept is on the one hand motivated by our remodeling approach presented in the following chapter, but on the other hand allows for more control in the interaction of the different layers. Admittedly, a higher numerical effort results and special element technology has to be implemented to account for the large aspect ratio.

4.4 Continuum Molecule Chain Models

The previously described material models follow the classical continuum mechanics approach to establish an invariant based macroscopic model governed by a number of abstract parameters. They try to phenomenologically model an observed material behavior within a regime of relevance. Though they rely on a profound theory of continuum mechanics they nevertheless lack a direct physical connection to the underlying mechanisms of deformation. Thus their governing parameters have to be defined by experiments and fitting to the observed material behavior. It is generally a goal for material models to reduce the number of parameters for a specific model and to relate parameters to a clear physical meaning. One step into that direction is a material model which takes the microstructure of the material into account. This approach seems beneficial for the modeling of arterial walls, because the microstructure of collagen and elastin fibers plays a decisive role in their structural characteristics. Moreover, experiments to determine material parameters, especially in vivo, are difficult and identification of further physical correlations could simplify the correct modeling of the wall.

The development of such microstructurally motivated material models was significantly driven by describing the large strain behavior of rubber materials. The corresponding rubber experiments of Treloar [310] for uniaxial tension, biaxial tension and shear are known as the quintessential rubber data. The observed S-shaped load versus stretch curve exhibited in uniaxial tension is captured by many models, however some of them fail in describing the response under different states of deformation without changing the model parameters (see Arruda and Boyce [10]).

Rubber, as well as typical soft biological tissue, consists of a complex three-dimensional network of long, randomly oriented polymer chains laterally attached to each other at occasional points along their lengths. Polymer chains in general have many configurations of almost equal energy. Perturbing this configuration from equilibrium generates entropic forces counteracting the perturbation. This is the basis for entropy-based elasticity. Since a long chain molecule may adopt a large number of different configurations, the treatment of each of them individually would require an enormous effort. For instance, the vulcanization process of rubber can be modeled with molecular-dynamic simulations together with advanced statistical methods like the Monte Carlo method, but such simulations necessitate an extreme computational effort (see Böl and Reese [39]).

Thus, so-called *statistical mechanics* as an alternative approach to describe long chain molecules gained considerable attention, a concept originally developed in the context of entropic rubber elasticity (see Kuhn and colleagues [188, 189], or Treloar [309]). In 1943 Treloar has derived the Neo-Hookean free energy function (see Section 4.2.1) with identifying $\mu = Nk\vartheta$ as product of the number of chains per reference volume N, Boltzmann's constant k, and the absolute temperature ϑ. We refer to the book of Treloar [311] for a detailed description on rubber elasticity and the underlying statistical mechanics methods (see also Böl [38]).

The principal idea of a microstructural constitutive model for such materials is to consider on the one hand the main load carrying constituent, the polymer chain, and on the other hand the cross-linking between the chain network in a homogenized sense. A number of different chain network models have been proposed over the past 60 years (see Treloar [311] and Flory [92]). The common feature of all these network models is a characteristic unit cell which is assumed to provide an adequate representation of the underlying macromolecular network structure. One can distinguish two categories, affine and non-affine models. The first non-affine model was the four chain tetrahedron by Flory and Rehner [93]. Arruda and Boyce [10, 46] proposed a non-affine eight-chain isotropic cube model. Comparison of different chain network models can also be found in Miehe *et al.* [215]. The eight-chain model has been adapted to orthotropy by Bischoff *et al.* [32, 33] and employed to model collagenous tissue by Garikipati and colleagues [107, 109], Kuhl *et al.* [184], Zhang *et al.* [340], and Rodríguez and colleagues [250, 249]. It is described in detail in the following.

4.4.1 Mechanical Response of a Single Fiber

The idea of continuum chain models is the homogenization of microstructural constituents. The main load carrying constituents of the arterial wall are the collagen fibers. The characteristic feature of collagen molecules is their long, stiff, triple-stranded helical structure. They can be characterized as freely jointed rigid links or by the so-called wormlike chain or Kratky-Porod model (see Kratky and Porod [177]). Both are entropy-based constitutive models for one molecule chain where the key kinematic variable that characterizes the conformation of the chain is the end-to-end length r. In the following, both models and their implications are compared (see Fig. 4.1).

Freely jointed chain model

A macromolecule can be modeled as n freely jointed bonds each of fixed length l. The direction in space of any bond is entirely random and bears no relation to the direction in space of any other link in the chain. Such a randomly jointed chain automatically excludes valence angle or other restrictions on the freedom of motion of neighboring bonds. The characteristic parameter is the *contour length* $L = nl$. In order to define the statistical properties of such a randomly jointed chain we consider one end of the chain to be fixed at the origin and the other end to move in a random manner throughout the available space. However, though the motion is random,

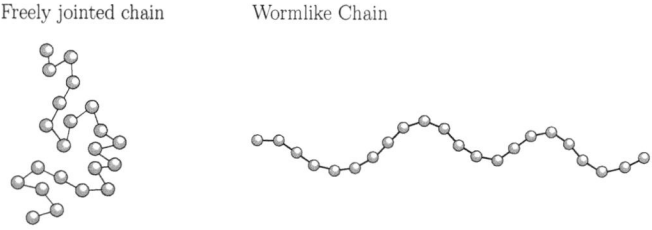

Fig. 4.1: Illustration of the two specified single chain models.

the probability for every end position is not equal. For any particular position $P(x,y,z)$ and its associated end-to-end length r there is an associated probability $p(x,y,z)\mathrm{d}v$ that the end shall be located within the volume element $\mathrm{d}v$ in the vicinity of the point P. According to the classical Boltzmann equation the configurational entropy of a single chain is proportional to the number of ways it can occupy space with its end at point P, given as

$$s = k \ln(p(x,y,z)\mathrm{d}v) \qquad (4.81)$$

with the Boltzmann constant $k = 1.380650E^{-23} J/K$. For purely entropic chains the free energy Ψ^{chn} of a single chain is $\Psi^{\mathrm{chn}} = -k\vartheta \ln(p(x,y,z)\mathrm{d}v)$ where ϑ is the absolute temperature.

It is a matter of the range of extension considered which statistical treatment is appropriate. If only moderate extensions are involved the Gaussian statistics treatment is sufficient. But as soon as the distance between the two end points of the chain approaches the fully extended length, Gaussian statistics becomes inadequate and more elaborate non-Gaussian treatment has to be used. For the classical Gaussian case the probability yields $p = p_0 \exp(-3/2Nr^2/L^2)$ with p_0 as parametric constant. The free energy of the individual chain takes the form

$$\Psi^{\mathrm{chn}}_{\mathrm{Gauss}} = \Psi^{\mathrm{chn}}_0 + \frac{3}{2}k\vartheta \frac{r^2}{nl^2}. \qquad (4.82)$$

The internal energy of the unperturbed state Ψ^{chn}_0 can usually be neglected (see Böl and Reese [39]). The necessary tensile force to elongate a chain in the direction of the line joining its ends follows straightforwardly as

$$f_{\mathrm{Gauss}} = \frac{\mathrm{d}\Psi^{\mathrm{chn}}_{\mathrm{Gauss}}}{\mathrm{d}r} = \frac{3k\vartheta}{l}\frac{r}{nl} \qquad (4.83)$$

being proportional to r. The linear force-extension derived from Gaussian statistics is limited to a distance r between the ends less than one-third of the fully extended length (see Treloar [311]).

For higher chain extensions a more accurate statistical distribution function has to be employed. Following Kuhn and Grün [189], non-Gaussian statistics of inverse Langevin type is well suited. The fractional extension of the chain $\frac{r}{nl}$ is described by the so-called Langevin function in terms of the parameter β

$$\frac{r}{nl} = \mathcal{L}(\beta) = \coth\beta - (1/\beta). \qquad (4.84)$$

To evaluate β the inverse Langevin function $\mathcal{L}^{-1}$ has to be evaluated. A series expansion yields

$$\beta = 3\frac{r}{nl} + \frac{9}{5}\left(\frac{r}{nl}\right)^3 + \frac{297}{175}\left(\frac{r}{nl}\right)^5 + \frac{1539}{875}\left(\frac{r}{nl}\right)^7 + \ldots . \qquad (4.85)$$

and the free energy of one single chain is obtained as

$$\Psi^{\text{chn}}_{\text{Langevin}} = \Psi^{\text{chn}}_0 + k\vartheta n \left(\frac{r}{nl}\beta + \ln\frac{\beta}{\sinh\beta}\right) \qquad (4.86)$$

where the internal energy Ψ^{chn}_0 can again be neglected and β is evaluated up to a specified accuracy according to (4.85). The tensile force is derived as in the Gaussian case by differentiation and yields together with the series expansion

$$f_{\text{Langevin}} = \frac{k\vartheta}{l}\mathcal{L}^{-1} = \frac{k\vartheta}{l}\left(3\frac{r}{nl} + \frac{9}{5}\left(\frac{r}{nl}\right)^3 + \frac{297}{175}\left(\frac{r}{nl}\right)^5 + \frac{1539}{875}\left(\frac{r}{nl}\right)^7 + \ldots\right). \qquad (4.87)$$

One can identify that the first term corresponds to the force of the Gaussian approximation (4.83).

Wormlike chain model

In contrast to the freely jointed chain models the wormlike chain model (WLC) considers a more continuous distribution of bond angles and therefore a certain relation between neighboring bonds. The model assumes the polymer chain as a flexible rod and is particularly suited to model stiffer polymers. It is characterized through a smooth curvature whose direction changes randomly but in a continuous manner (see Kratky and Porod [177] and Flory [92]). In addition to the contour length $L = nl$ the behavior of one chain is controlled by the so-called *persistence length* A. It is defined by the sum of the average projections of all bonds onto the direction of the first bond. It varies between $l \leq A \leq L$ and is a measure of the initial chain stiffness. The persistence length of the uncorrelated freely jointed chain would be equal to the length of the individual bond $A = l$ whereas the persistence length of an infinitely stiff chain with beamlike properties is equal to the contour length $A = L$. Several biologically important polymers have been modeled as wormlike chains such as DNA (see Marko and Siggia [207] and Bustamante et al. [50]) and collagen (see Bischoff et al. [33]).

The strain energy of the wormlike chain model is given as

$$\Psi^{\text{chn}}_{\text{wlc}} = \Psi^{\text{chn}}_0 + k\vartheta\frac{L}{4A}\left(2\frac{r^2}{L^2} + \frac{1}{1-\frac{r}{L}} - \frac{r}{L}\right). \qquad (4.88)$$

Herein, as before, Ψ^{chn}_0 is the energy of the chain at the undeformed state, k the Boltzmann constant, ϑ the absolute temperature, and r the end-to-end length of the chain. It is derived by integrating the force-stretch relation for the wormlike chain

$$f_{\text{wlc}} = \frac{k\vartheta}{4A}\left(4\frac{r}{L} + \frac{1}{\left(1-\frac{r}{L}\right)^2} - 1\right) \qquad (4.89)$$

proposed by Marko and Siggia [207].

Comparison of different chain models

To illustrate the behavior of the two different single chain models, the freely jointed chain model with Gaussian and Langevin statistics and the wormlike chain model (see also Fig. 4.1), some force-elongation curves are compared in Fig. 4.2. The different chain forces f are hereby scaled by $\frac{1}{k\vartheta}$ and plotted versus the chain stretch $\frac{r}{L}$. The freely jointed chain curves are plotted for unit bond length $l = 1$. The linear behavior of the chain with Gaussian statistics is clearly observed, whereas the nonlinear curve of the chain with Langevin statistics represents the same initial slope with strong stiffening for large extensions reaching the chain contour length. The wormlike chain model thus reflects the highly nonlinear behavior with the characteristic locking effect when approaching the chain contour length. In addition, the second parameter, the persistence length A, controls the curvature and thus the rate of the rising force when the locking stretch is reached. For smaller values of A reflecting highly curved chains the force response is smoother, because the chain needs to be bent up.

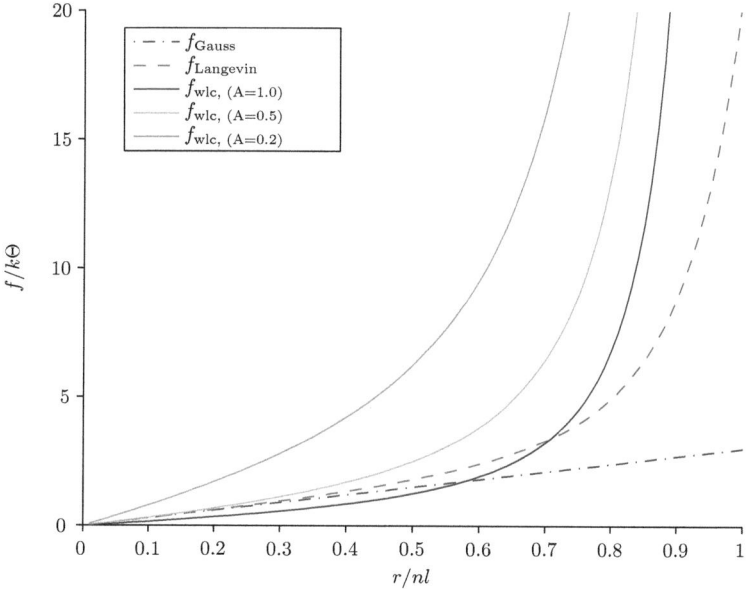

Fig. 4.2: Force-elongation curves for different single chain models.

4.4.2 Chain Network Model

On the macroscopic scale the network of the individual chains needs to be modeled. The use of Gaussian and Langevin statistics in networks has been considered by several authors (see for instance Flory and Rehner [93], Wang and Guth [328] and Treloar [311]). Arruda and Boyce [10] propose an eight-chain cubic unit cell as sketched in Fig. 4.3. The model is extended to

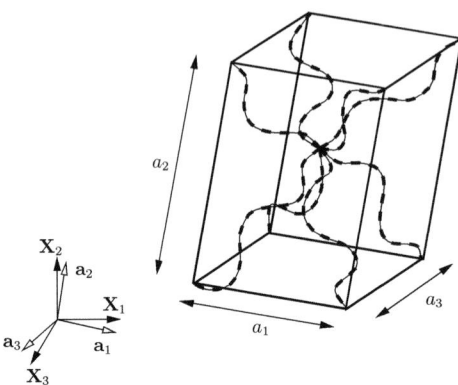

Fig. 4.3: Eight-chain orthotropic unit-cell with cell dimensions a_i and material axes $\mathbf{a}_i$ with respect to coordinate system $\mathbf{X}_i$.

orthotropic behavior by Bischoff et al. [32, 33]. The unit cell is oriented in space such that the principal stretches are applied along fixed cell directions. Where equal cell dimensions insure isotropy of the initial mechanical response with respect to the principal stretch space, initial orthotropy of a network with preferred fiber orientation can be modeled by different cell dimensions a_i, $i = 1, 2, 3$ (see Fig. 4.3). The orthotropy results from a fixed orientation of the unit cell in the space specified by the orthogonal principal material axes $\mathbf{a}_i$ with respect to a reference coordinate system $\mathbf{X}_i$ together with different cell dimensions a_i along these axes.

The altogether eight chains within one unit cell have all the same undeformed chain length $r_0 = \sqrt{a_1^2 + a_2^2 + a_3^2} = \sqrt{a_i^2}$. Note that this implicitly defines the underlying chain contour length $L = nl$ which is usually taken to be the root mean square length. An affine deformation, that is the ends of the chains are fixed in the continuum and deform with the continuum strain field described by $\boldsymbol{F}$ (or $\boldsymbol{C}$), is assumed. The end-to-end length r of the chains in the deformed configuration is evaluated as

$$r = \sqrt{a_i^2\, \mathbf{a}_i \cdot \boldsymbol{C} \cdot \mathbf{a}_i}. \tag{4.90}$$

We can define the non-standard invariants $\bar{I}_i$ as $\bar{I}_i = \mathbf{a}_i \cdot \boldsymbol{C} \cdot \mathbf{a}_i$ following the notation for anisotropic tensor functions in Section 4.1.4. They represent the squared stretch in each unit cell direction i. The deformed lengths are functions of these invariants and thus satisfy the appropriate symmetry requirements (Bischoff et al. [32]). To employ the wormlike chain model for one single chain within the continuum network model a chain number density N is introduced scaling the contributions of the eight individual chains with the number of chains per unit volume. This results in a free energy of a representative volume element of

$$\Psi^{\text{chn}} = k\vartheta \frac{NL}{4A} \left(2\frac{r^2}{L^2} + \frac{1}{1-\frac{r}{L}} - \frac{r}{L} \right). \tag{4.91}$$

4. Constitutive Models for the Arterial Wall

To achieve a stress-free undeformed state an additional repulsive energy is introduced

$$\Psi^{\text{rep}} = -k\vartheta \frac{NL}{4A} \left(\frac{1}{L} + \frac{1}{4r_0(1-\frac{r_0}{L})^2} - \frac{1}{4r_o} \right) \ln(\bar{I}_1^{2a_1^2} \bar{I}_2^{2a_2^2} \bar{I}_3^{2a_3^2}). \tag{4.92}$$

This term reflects a mutual repulsion of chains from each other that will prevent the entropic collapse of the unit cell while maintaining the orthotropic shape of the unit cell. Considering that the invariants $\bar{I}_i$ represent the stretches along the principal directions the term $\ln(\bar{I}_1^{2a_1^2} \bar{I}_2^{2a_2^2} \bar{I}_3^{2a_3^2})$ is equivalent to $\ln(\lambda_{\mathbf{a}_1}^{2a_1^2} \lambda_{\mathbf{a}_2}^{2a_2^2} \lambda_{\mathbf{a}_3}^{2a_3^2})$ in terms of the strains $\lambda_{\mathbf{a}_i}$ in each cell direction $\mathbf{a}_i$. A similar term is employed within the repulsive energy proposed by Bischoff et al. [32]. Note that this definition differs from the formulation proposed by Kuhl and Holzapfel [185], who introduce the repulsive energy term $\ln(\bar{I}_1 \bar{I}_2 \bar{I}_3)$ which does not satisfy the stress-free reference state.

To allow for control of compressibility of the material and eventually model isotropic ground substance an additional bulk contribution is added to the energy of the volume element. It is chosen as standard Neo-Hookean type in terms of the Lamé parameters Λ, μ

$$\Psi^{\text{blk}} = \frac{1}{2}\Lambda \ln^2(J) + \frac{1}{2}\mu(I_1 - 3) - \mu \ln(J). \tag{4.93}$$

The final strain energy of the material based on the orthotropic cube network model is the sum of the former components

$$\Psi^{\text{ChainNetw}} = \Psi^{\text{blk}} + \Psi^{\text{chn}} + \Psi^{\text{rep}}. \tag{4.94}$$

Depending of the cell dimensions, isotropic, transverse isotropic and orthotropic material behavior can be modeled.

Stress and elasticity tensor

The second Piola-Kirchhoff stress is derived from $\boldsymbol{S} = 2\frac{d\Psi}{d\boldsymbol{C}}$ to the following form

$$\boldsymbol{S}^{\text{ChainNetw}} = 2\frac{\partial \Psi^{\text{blk}}}{\partial \boldsymbol{C}} + 2\frac{\partial \Psi^{\text{chn}}}{\partial \boldsymbol{C}} + 2\frac{\partial \Psi^{\text{rep}}}{\partial \boldsymbol{C}}$$
$$= \boldsymbol{S}^{\text{blk}} + \boldsymbol{S}^{\text{chn}} + \boldsymbol{S}^{\text{rep}}, \tag{4.95}$$

$$\boldsymbol{S}^{\text{blk}} = (\Lambda \ln(J) - \mu)\boldsymbol{C}^{-1} + \mu \boldsymbol{I}, \tag{4.96}$$

$$\boldsymbol{S}^{\text{chn}} = \frac{k\vartheta N}{4A} \left(\frac{1}{L} + \frac{1}{4r(1-\frac{r}{L})^2} - \frac{1}{4r} \right) a_i^2 \mathbf{a}_i \otimes \mathbf{a}_i, \tag{4.97}$$

$$\boldsymbol{S}^{\text{rep}} = -\frac{k\vartheta N}{4A} \left(\frac{1}{L} + \frac{1}{4r_0(1-\frac{r_0}{L})^2} - \frac{1}{4r_0} \right) \frac{4a_i^2}{\bar{I}_i} \mathbf{a}_i \otimes \mathbf{a}_i. \tag{4.98}$$

We introduce the structural tensors $\boldsymbol{A}_i = \mathbf{a}_i \otimes \mathbf{a}_i$ with $\text{tr}[\boldsymbol{A}_i] = 1$ for each cell direction. The elasticity tensor follows as

$$\mathbb{C}^{\text{ChainNetw}} = 2\frac{\partial \boldsymbol{S}^{\text{blk}}}{\partial \boldsymbol{C}} + 2\frac{\partial \boldsymbol{S}^{\text{chn}}}{\partial \boldsymbol{C}} + 2\frac{\partial \boldsymbol{S}^{\text{rep}}}{\partial \boldsymbol{C}}$$
$$= \mathbb{C}^{\text{blk}} + \mathbb{C}^{\text{chn}} + \mathbb{C}^{\text{rep}}, \tag{4.99}$$

$$\mathbb{C}^{\text{blk}} = \Lambda \boldsymbol{C}^{-1} \otimes \boldsymbol{C}^{-1} + 2(\mu - \lambda \ln(J))\boldsymbol{C}^{-1} \boxtimes \boldsymbol{C}^{-1}, \tag{4.100}$$

$$\mathbb{C}^{\text{chn}} = \frac{k\vartheta N}{4A}\frac{1}{r^3}\left(1 - \frac{1}{(1-\frac{r}{L})^2} + \frac{2r}{L(1-\frac{r}{L})^3}\right) a_i^4 \boldsymbol{A}_i \otimes \boldsymbol{A}_i, \tag{4.101}$$

$$\mathbb{C}^{\text{rep}} = \frac{k\vartheta N}{4A}\left(\frac{1}{L} + \frac{1}{4r_0(1-\frac{r_0}{L})^2} - \frac{1}{4r_0}\right) \frac{8a_i^2}{I_i^2} \boldsymbol{A}_i \otimes \boldsymbol{A}_i. \tag{4.102}$$

This material model is characterized by its underlying micro- and macroscopic behavior, namely the molecule chain behavior and the network model. Together with the ground substance it is well suited to model biomechanical material, especially arterial walls with collagen chains as main load carrying constituents. The necessary parameters are λ and μ for the ground substance, the chain characteristics L, A and the chain number density N which are motivated micromechanically, and the three cell dimension a_i where the relation between the undeformed chain length $r_0 = \sqrt{a_i}$ should be respected. Typically, a micromechanically motivated value r_0 is assumed and the cell dimensions are adapted accordingly. This set of parameters needs to be obtained by experiments and curve fitting procedures. Moreover, such a formulation is especially appealing for material remodeling approaches. In contrast to the cumbersome handling of vectors to describe the direction of anisotropy, the cell dimensions are easier to manipulate numerically. Therefore, we employ the presented chain network constitutive law within a proposed remodeling procedure, described in Chapter 5.

4.5 Extension to Viscoelasticity

The mechanical behavior of most biological soft tissue is in general nonlinear viscoelastic rather than hyperelastic. Although arterial walls are typically characterized by a hysteresis loop these are usually considered independent on the rate of strain, as proposed among others by Fung [102]. However, several studies have proven that a certain stage of viscoelasticity is present also in arterial walls and we refer to the reviews of Haslach [127] and Kalita and Schaefer [164] for detailed information. Obviously also in other cardiovascular problems viscoelasticity plays a significant role, for instance the thrombus within aneurysms features viscoelasticity as examined by Dam et al. [314].

A number of viscoelastic formulations have been proposed in order to extend the well known linear rheological models to the large strain regime. Lubliner [201] proposed a split of the free energy of a viscoelastic solid in two parts: the first part describing the rate-independent material behavior and the second incorporating time-dependent effects. He further assumed a multiplicative decomposition of the deformation gradient into elastic and inelastic parts with inelastic strain.

4. Constitutive Models for the Arterial Wall

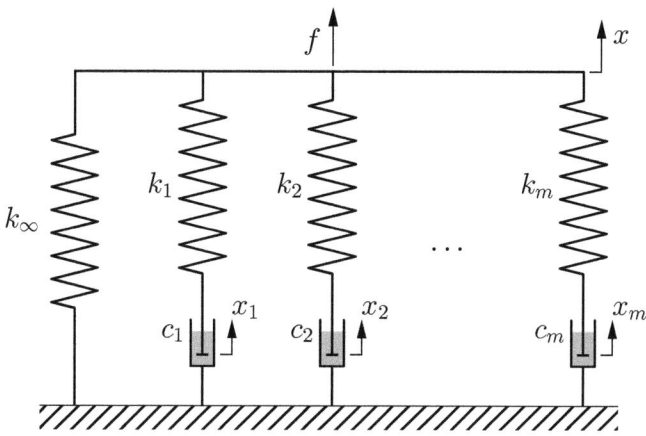

Fig. 4.4: Generalized Maxwell-Model.

A set of conjugate pairs of state and control variables together with a generalized energy function relating one with the other ensure thermodynamical consistency. More recently, a strategy to build a viscoelastic model has been to write the Helmholtz energy as the sum of a long-term hyperelastic and a viscous energy (see for instance Bonet [42]). Either the over-stress or the inelastic strain can be chosen as internal variable driving the dissipation process. One of the first finite element implementations for a finite linear viscoelasticity was given 1987 by Simo [263]. He chose the over-stress as the internal variable and this approach was followed by Govindjee and Simo [115] and Holzapfel [136] (see also Reese and Govindjee [244]).

An important point in developing models of this form is the choice of the evolution equations for the internal variables. In the theory of linear viscoelasticity which is only valid for small deformations and small perturbations away from thermodynamic equilibrium, the question is of minor importance as the relationship between stress and strain is linear. In the case of large deformations, however, the choice of internal variables and evolution equations is not so evident and not unique. In the following we assume only small deviations away from thermodynamic equilibrium and will speak of finite linear viscoelasticity. Note that finite deformations are still valid. Evolution equations belonging to this kind of theory are linear differential equations with possibly non-constant coefficients, for instance a deformation-dependent relaxation time.

4.5.1 Linear Generalized Maxwell-Model

To clarify the concept of viscoelasticity we present the rheological model of a one-dimensional so-called Maxwell-body employing linear (small strain) theory (see Fig. 4.4). It is a popular model to formulate time-dependent and frequency dependent characteristics and is composed of an elastic spring with stiffness k_∞ in parallel with m so-called Maxwell-elements. Each element i consists of a spring k_i in series with a dashpot c_i.

The total force acting on the model is composed of a long term or steady state component f_∞ plus an arbitrary number m of so-called nonequilibrium forces f_i in the sense of non-equilibrium thermodynamics

$$f = f_\infty + \sum_{\alpha=1}^{m} f_\alpha \tag{4.103}$$

Here, and from now on $(\,\cdot\,)_\infty$ characterizes a quantity related to a sufficiently slow and therefore purely elastic process. Introducing the constitutive equations for the springs, (4.103) yields

$$f = k_\infty x + \sum_{\alpha=1}^{m} k_\alpha (x - x_\alpha) \tag{4.104}$$

where the displacements x_α are unknown *internal variables*. They are defined by the equilibrium relationship between each spring and dashpot of one Maxwell-element

$$c_\alpha \dot{x}_\alpha = f_\alpha$$
$$f_\alpha = k_\alpha (x - x_\alpha) \tag{4.105}$$

where the dashpot constant c_α represents its linear viscosity and can be related to the stiffness of the corresponding spring in terms of a *retardation time*, or *relaxation time* parameter τ_α as

$$c_\alpha = \tau_\alpha k_\alpha. \tag{4.106}$$

This allows to rewrite (4.105) as evolution equation for the internal variable x_α

$$\dot{x}_\alpha = \frac{1}{\tau_\alpha}(x - x_\alpha) \tag{4.107}$$

relating strain and strain rate. Alternatively, time differentiation of (4.105) yields

$$\dot{f}_\alpha + \frac{1}{\tau_\alpha} f_\alpha = k_\alpha \dot{x} \tag{4.108}$$

as evolution equation in terms of force and force rate. Such a first-order differential equation has a closed form solution

$$f_\alpha = \exp(-T/\tau_\alpha)\left. f_\alpha \right|_{t=0} + \int_{t=0}^{t=T} \exp\left(-\frac{T-t}{\tau_\alpha}\right) \dot{f}_\alpha(t) dt, \quad \alpha = 1, \ldots, m \tag{4.109}$$

Now we introduce the energy terms of the above problem in order to generalize it to the large strain three-dimensional case. The total free energy of the system is

$$\Psi(x, x_1, \ldots, x_m) = \Psi_\infty(x) + \sum_{\alpha=1}^{m} \Psi_\alpha(x, x_\alpha) = \tfrac{1}{2} k_\infty x^2 + \sum_{\alpha=1}^{m} \tfrac{1}{2} k_\alpha (x - x_\alpha)^2. \tag{4.110}$$

The rate of work dissipated within each dashpot yields the total internal dissipation

$$\mathcal{D}_{\text{int}} = \sum_{\alpha=1}^{m} f_\alpha \dot{x}_\alpha = \sum_{\alpha=1}^{m} c_\alpha (\dot{x}_\alpha)^2 \geq 0 \tag{4.111}$$

which is always non-negative and disappears at equilibrium. Note that differentiation of the total free energy $\Psi(x, x_1, \ldots, x_m)$ with respect to the total strain x yields equation (4.104) and with respect to the internal variables x_α yields

$$\frac{\partial \Psi(x, x_1, \ldots, x_m)}{\partial x_\alpha} = -k_\alpha(x - x_\alpha) = -f_\alpha, \qquad \alpha = 1, \ldots, m \qquad (4.112)$$

and the internal dissipation can be expressed through the strain energy

$$\mathcal{D}_{\text{int}} = -\sum_{\alpha=1}^{m} \frac{\partial \Psi(x, x_1, \ldots, x_m)}{\partial x_\alpha} \dot{x}_\alpha. \qquad (4.113)$$

These relationships will be used in the generalization to the three-dimensional large strain regime described in the following section.

4.5.2 Large Strain Fiber-Reinforced Viscoelasticity

In the section at hand we discuss an anisotropic visco-hyperelastic material model following closely the work of Holzapfel and co-workers [136, 137, 139, 141] (see also Kaliske [163]), which is derived from the generalization of the linear one-dimensional visco-elastic model discussed in the previous section.

The concept of internal variables is well established for modeling viscoelasticity. They are not accessible to direct observation, but are used to describe the material behavior associated with irreversible dissipative effects. So far, $\boldsymbol{C}$ representing the strain state has always been employed as driving parameter for hyperelastic materials. Accordingly, we introduce a number of internal *strain-like* control variables describing the viscous behavior, each identified by a tensor $\boldsymbol{\Gamma}_\alpha$, $\alpha = 1, \ldots, m$. Equivalently, the viscoelastic state is assumed to depend on internal stress-like variables, each described by a tensor $\boldsymbol{Q}_\alpha$, $\alpha = 1, \ldots, m$. Each conjugate pair of internal variables α is related to one viscoelastic process with its relaxation time $\tau_\alpha \in (0, \infty)$. Note that the viscoelastic response of arterial walls is characterized by a almost constant damping over a wide, continuous frequency spectrum, which would need an infinite number m. However, often the complex viscoelastic behavior is adequately modeled by a finite number of m internal state variables.

According to the free energy of the linear Maxwell model we define the Helmholtz free energy function per unit reference volume of the viscoelastic continuous body as

$$\Psi^{\text{visco}} = \Psi(\boldsymbol{C}, \boldsymbol{\Gamma}_1, \ldots, \boldsymbol{\Gamma}_m). \qquad (4.114)$$

As we want to model viscoelasticity of anisotropic fiber-reinforced materials we relate the elastic response to the hyperelastic Holzapfel-model, see Section 4.3.1. Thus, a further *decoupled* representation reads

$$\begin{aligned}
\Psi^{\text{visc-ani}} &= \Psi_\infty^{\text{vol}}(J) + \Psi^{\text{visco}}(\widehat{\boldsymbol{C}}, \boldsymbol{M}, \boldsymbol{N}, \boldsymbol{\Gamma}_1, \ldots, \boldsymbol{\Gamma}_m) \\
&= \Psi_\infty^{\text{vol}}(J) + \Psi_\infty^{\text{aniso}}(\widehat{I}_1, \widehat{I}_2, \widehat{I}_4, \widehat{I}_6) + \sum_{\alpha=1}^{m} \Upsilon_\alpha(\widehat{\boldsymbol{C}}, \boldsymbol{M}, \boldsymbol{N}, \boldsymbol{\Gamma}_\alpha)
\end{aligned} \qquad (4.115)$$

where we assume that viscous effects are only related to isochoric deformation. We hereby introduce scalar-valued functions Υ_α, $\alpha = 1, \ldots, m$ as so-called configurational free energies and their sum as 'dissipative potential'.

Following the concept described in the previous section we define

$$\boldsymbol{S} = 2\frac{\partial \Psi(\boldsymbol{C}, \boldsymbol{M}, \boldsymbol{N}, \boldsymbol{\Gamma}_1, \ldots, \boldsymbol{\Gamma}_m)}{\partial \boldsymbol{C}} = \boldsymbol{S}_\infty^{\text{vol}} + \boldsymbol{S}_\infty^{\text{blk}} + \boldsymbol{S}_\infty^{\text{fib}} + \sum_{\alpha=1}^{m} \boldsymbol{Q}_\alpha \qquad (4.116)$$

for the second Piola-Kirchhoff stress. We thus obtain the conjugate *non-equilibrium stresses* $\boldsymbol{Q}_\alpha$ as

$$\boldsymbol{Q}_\alpha = 2\frac{\partial \Upsilon_\alpha(\widehat{\boldsymbol{C}}, \boldsymbol{M}, \boldsymbol{N}, \boldsymbol{\Gamma}_\alpha)}{\partial \boldsymbol{C}}, \qquad \alpha = 1, \ldots, m \qquad (4.117)$$

and within the isochore-volumetric decoupling and the deviatoric projection, referring to (4.24),

$$\boldsymbol{Q}_\alpha = J^{-2/3}\mathbb{P} : \widehat{\boldsymbol{Q}}_\alpha, \qquad \widehat{\boldsymbol{Q}}_\alpha = 2\frac{\partial \Upsilon_\alpha(\widehat{\boldsymbol{C}}, \boldsymbol{M}, \boldsymbol{N}, \boldsymbol{\Gamma}_\alpha)}{\partial \widehat{\boldsymbol{C}}}, \qquad (4.118)$$

we define the *fictitious non-equilibrium stresses* $\widehat{\boldsymbol{Q}}_\alpha$.

According to (4.113) the non-negative internal dissipation is obtained as

$$\mathcal{D}_{\text{int}} = -\sum_{\alpha=1}^{m} 2\frac{\partial \Upsilon_\alpha(\widehat{\boldsymbol{C}}, \boldsymbol{M}, \boldsymbol{N}, \boldsymbol{\Gamma}_\alpha)}{\partial \boldsymbol{\Gamma}_\alpha} : \frac{1}{2}\dot{\boldsymbol{\Gamma}}_\alpha \geq 0 \qquad (4.119)$$

and the conjugate pairs $\boldsymbol{Q}_\alpha$ and $\boldsymbol{\Gamma}_\alpha$ are related via the internal constitutive equations

$$\boldsymbol{Q}_\alpha = -2\frac{\partial \Upsilon_\alpha(\widehat{\boldsymbol{C}}, \boldsymbol{M}, \boldsymbol{N}, \boldsymbol{\Gamma}_\alpha)}{\partial \boldsymbol{\Gamma}_\alpha}, \qquad \alpha = 1, \ldots, m. \qquad (4.120)$$

This relation thereby restricts the functions Υ_α to satisfy conditions (4.117) and (4.120). The condition for thermodynamic equilibrium implies that the stress reaches equilibrium for $t \to \infty$, which means that $\lim_{t \to \infty} \boldsymbol{Q}_\alpha = \boldsymbol{0}$.

Evolution equations and their algorithmic implementation

In order to fully determine how a viscoelastic process evolves we have to postulate additional equations of evolution governing the internal variables, in our case the non-equilibrium stresses $\boldsymbol{Q}_\alpha$. They should provide a good approximation to the observed physical behavior and be suitable for efficient time integration algorithms within a finite element framework. As mentioned earlier we model finite linear viscoelasticity assuming linear evolution equations which have to satisfy inequality (4.119).

Following the idea of Holzapfel and colleagues we formulate an evolution equation *separately* for each isochoric stress contribution $\boldsymbol{S}_\infty^{\text{iso}\,a}$ and each corresponding viscoelastic process α. In the case of the Holzapfel-model Ψ^{HGO} we have $\boldsymbol{S}_\infty^{\text{iso}\,a} \in \{\boldsymbol{S}_\infty^{\text{blk}}, \boldsymbol{S}_\infty^{\text{fib},1}, \boldsymbol{S}_\infty^{\text{fib},2}\}$. According to (4.108) a possible set of linear differential equations is

$$\dot{\boldsymbol{Q}}_\alpha^a + \frac{1}{\tau_\alpha^a}\boldsymbol{Q}_\alpha^a = \beta_\alpha^a \dot{\boldsymbol{S}}_\infty^{\text{iso}\,a}, \qquad \boldsymbol{Q}_\alpha^a\big|_{t=0} = \boldsymbol{0} \qquad (4.121)$$

valid for some semi-open time interval $t \in (0, T]$. The initial conditions ensure that the reference configuration has no viscoelastic stress contribution. The parameters $\beta_\alpha^a \in [0, \infty)$ are non-dimensional so-called *free-energy factors* associated with the relaxation times $\tau_\alpha^a \in (0, \infty)$, which describe the rate of decay of the stress and strain in a viscoelastic process.

Closed-form solutions according to (4.109) may be represented by simple convolution integrals

$$Q_\alpha^a = \int_{t=0^+}^{t=T} \exp\left(-\frac{T-t}{\tau_\alpha^a}\right) \beta_\alpha^a \dot{S}_\infty^{\text{iso}\,a}(t)\, dt \qquad (4.122)$$

Within a finite element framework a suitable numerical integration scheme is applied where we choose a one-step-θ algorithm with θ as algorithmic parameter, see for instance Zienkiewicz and Taylor [341] for details. The necessary algorithmic stress and elasticity tensors at time t_{n+1} are derived in the following.

We consider a certain time sub-interval $[t_n, t_{n+1}]$ with $\Delta t_n = t_{n+1} - t_n$ representing the associated time increment. Assuming that all relevant kinematic quantities are given at t_n as well as at t_{n+1} and also the stress S_n is known serving as history data. It then remains to specify at t_{n+1} the *algorithmic stress tensor*

$$S_{n+1} = \left[S_\infty^{\text{vol}} + S_\infty^{\text{fib}} + \sum_{\alpha=1}^m Q_\alpha\right]_{n+1}. \qquad (4.123)$$

The first two contributions $S_{\infty\,n+1}^{\text{vol}}$ and $S_{\infty\,n+1}^{\text{fib}}$ are directly available from the given strain measures at t_{n+1}, but the third term depends on the convolution integral (4.122). For each contribution a and α we obtain by applying the one-step-θ algorithm

$$Q_{\alpha\,n+1}^a = \underbrace{\frac{1}{1 + \frac{\theta \Delta t}{\tau_\alpha^a}} \left(\left(1 - \frac{(1-\theta)\Delta t}{\tau_\alpha^a}\right) Q_{\alpha\,n}^a - \beta_\alpha^a S_{\infty\,n}^{\text{iso}\,a}\right)}_{\mathcal{H}_{\alpha n}^a} + \frac{1}{1 + \frac{\theta \Delta t}{\tau_\alpha^a}} \beta_\alpha^a S_{\infty\,n+1}^{\text{iso}\,a} \qquad (4.124)$$

where $\mathcal{H}_{\alpha n}^a$ represents a history term for each contribution a, α. Accordingly, the *algorithmic elasticity tensor* at t_{n+1} is obtained as

$$\mathbb{C}_{n+1} = \left[\mathbb{C}_\infty^{\text{vol}} + \sum_a \left(1 + \frac{1}{1 + \frac{\theta \Delta t}{\tau_\alpha^a}} \beta_\alpha^a\right) \mathbb{C}_\infty^{\text{iso}\,a}\right]_{n+1}. \qquad (4.125)$$

The present constitutive law allows the modeling of viscoelastic fiber-reinforced materials at finite strains. To control viscoelasticity, for every contribution a and α two additional parameters τ_α^a and β_α^a need to be specified. The relaxation time governs the rate of decay of the viscoelastic stress contributions and the non-dimensional *free-energy factors* relate the stiffness of the viscoelastic process to the purely elastic process. These have to be determined experimentally.

4.6 Numerical Examples

4.6.1 Stretching of a Rubber Sheet

With this simulation of a rubber sheet we demonstrate the ability to model incompressible material behavior within large strains and deformations. The example has been analyzed by Klinkel et al. [172] and serves here as a benchmark for the proposed Mooney-Rivlin-type material law (see Section 4.2.2).

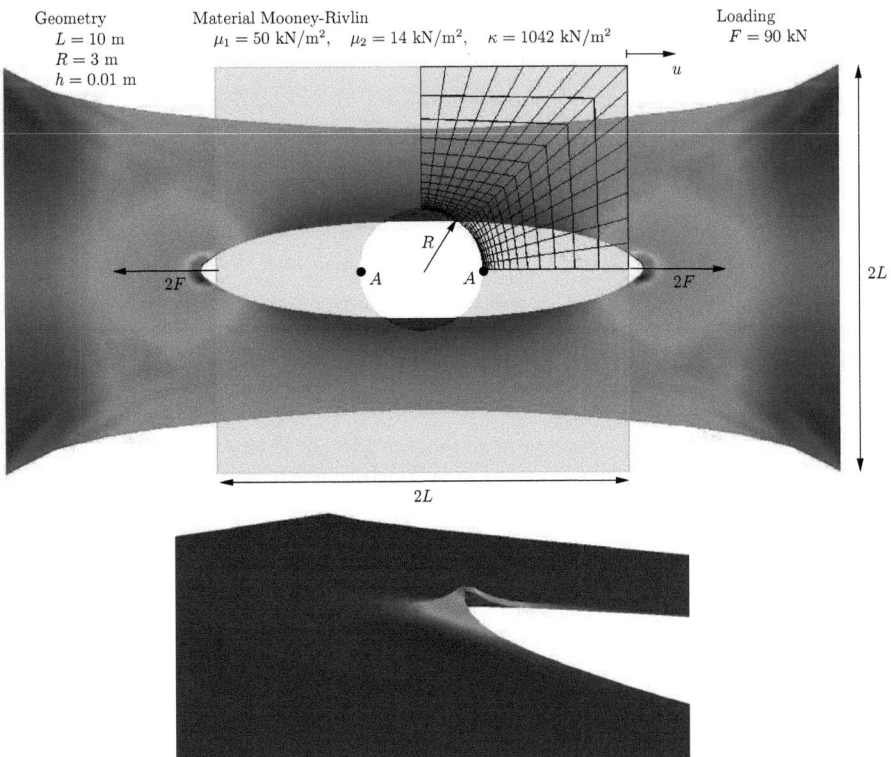

Fig. 4.5: Stretching of a rubber sheet. Initial and deformed structure (top) with colors representing the thickness-stretch. The mesh of a quarter system consists of 16 by 16 `Sosh8`-elements. A close-up illustrates the out-of-plane deformation (bottom).

A square sheet with a hole is stretched up to twice its original length, where we simulate a quarter of the system and apply appropriate symmetry conditions. The rubber material is modeled with Mooney-Rivlin-type materials where we compare the results of the specific strain energy functions $\Psi^{\text{MR,Klinkel}}$ (4.60) and Ψ^{MR} (4.61). The problem is depicted in Fig. 4.5 together with the final deformation state. The displacements at both ends are fixed in vertical direction and constraint to the same horizontal displacement. At the inner edge of the hole

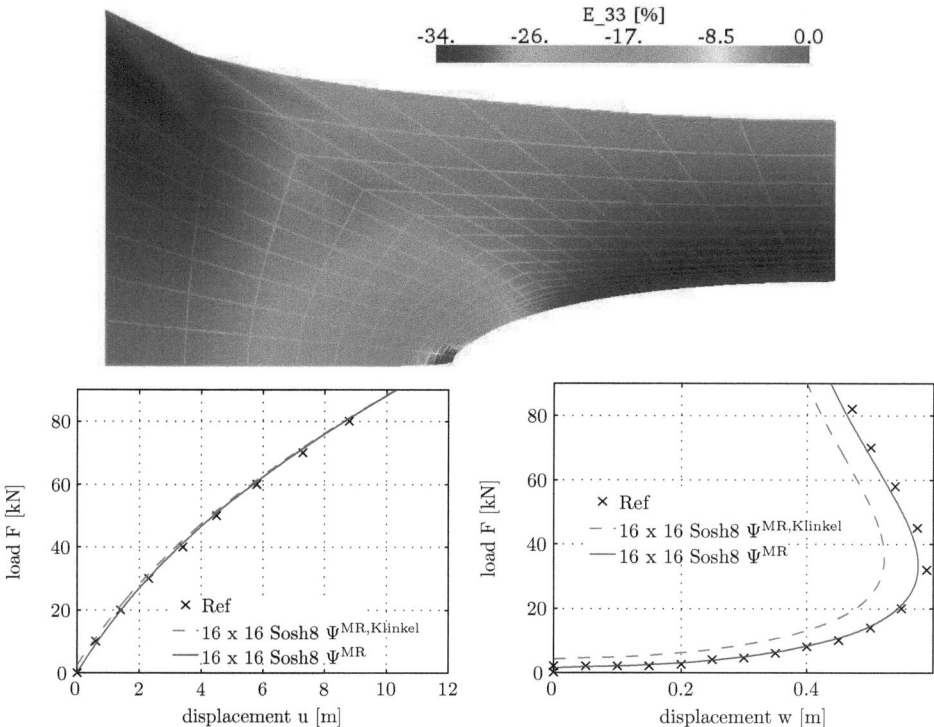

Fig. 4.6: Results for the stretched rubber sheet. The Green-Lagrangean thickness strain (top) recovers well the literature results. The load-deflection diagrams for the horizontal displacement (left) and out-of-plane displacement (right) point out the difference of a strain-energy with (Ψ^{MR}) and without ($\Psi^{MR,Klinkel}$) a stress-free reference configuration.

a stability problem is observed. A perturbation load in direction orthogonal to the sheet is applied at points A with $\tilde{F} = 10^{-7}F$ to follow the secondary equilibrium path characterized by an out-of-plane deflection (see also the close-up in Fig. 4.5).

The results are depicted in Fig. 4.6. As reference solution we refer to the paper by Klinkel et al. [172]. The depicted thickness strain correlates well with the reference plot as well as the load-displacement curve for the horizontal displacement of the edge. However, the results for the out-of-plane displacement w at the perturbation points A are worth mentioning. Where the results with the strain-energy function $\Psi^{MR,Klinkel}$ reported in the paper are slightly off the reference curve, our proposed modified strain-energy Ψ^{MR} yields almost exactly the reference curve. The difference is thus related to the fact that $\Psi^{MR,Klinkel}$ is not stress-free in reference configuration whose significance is therefore demonstrated.

4.6.2 Plate with Differing Warp and Fill

This example is taken from Balzani et al. [18] to illustrate the effect of anisotropic materials. A typical membrane-like engineering material is characterized by different behavior in warp- and fill-direction. A thin quadratic plate with such a material is subject to an orthogonal pressure load. Following the reference paper where the plate is discretized with two-dimensional shell elements we apply a Navier-support by fixing the lower edges of our three-dimensional Sosh8-discretization with 20 by 20 elements. The anisotropic material law proposed by Balzani and colleagues and described in Section 4.3.2 is applied with two different parameter sets for the warp- and fill-direction. The problem and the material parameters are presented in Fig. 4.7.

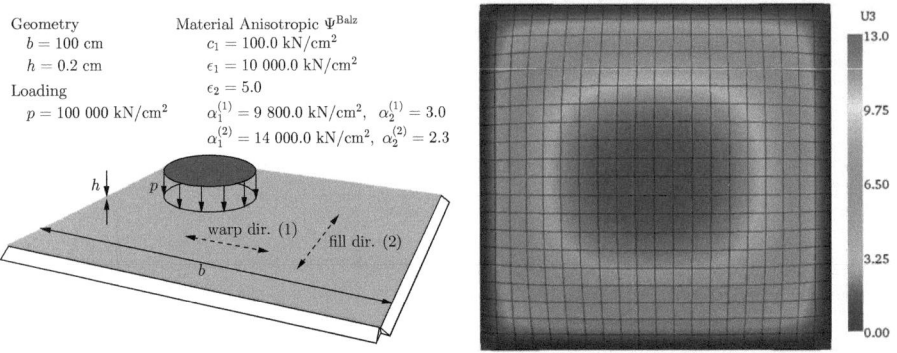

Fig. 4.7: Quadratic Navier-supported plate subject to orthogonal pressure with anisotropic material Ψ^{Balz} with differing warp and fill. Out-of-plane displacement shows twofold symmetry due to different warp- and fill-stiffness.

The result resembles nicely the ones presented in the reference. The out-of-plane displacement depicted on the right of Fig. 4.7 is characterized by the different stiffness in warp- and fill-direction. Thus, instead of a fourfold symmetric displacement expected for an isotropic material the twofold symmetry of the material becomes obvious from the oval displaced shape. Note that the implementation of such a three-dimensional material law into a framework with advanced solid-shell-elements is straight-forward and a special handling of the thickness-direction as discussed in Balzani et al. [18] is obsolete.

4.6.3 Inflation of a Fiber-Reinforced Rubber Tube

We consider the elastic response of a long circular tube continuously reinforced by two families of fibers symmetrically wound in helical manner along the axial direction (see Fig. 4.8). Structural elements of this type are frequently employed in industrial applications, but are also often found in biomechanical problems, like blood vessels for example. Holzapfel and Gasser [139] and Nedjar [218] have dealt with such problems from a computational mechanics point of view and serve here as reference basis.

We consider a 200 mm long portion of a tube with inner radius of 100 mm and thickness of 5 mm. One quarter of the structure is discretized with 10 × 20 × 2 Sosh8-elements and symmetry boundary conditions are applied at the lower and side surfaces. The tube is subject to an increasing internal pressure p which results also in an increasing axial force $F(p, r_i) = p\pi r_i^2$ which depends on the current internal radius r_i of the deformed structure, see also the sketch on the right of Fig. 4.8. This force is distributed to the top surface of the structure.

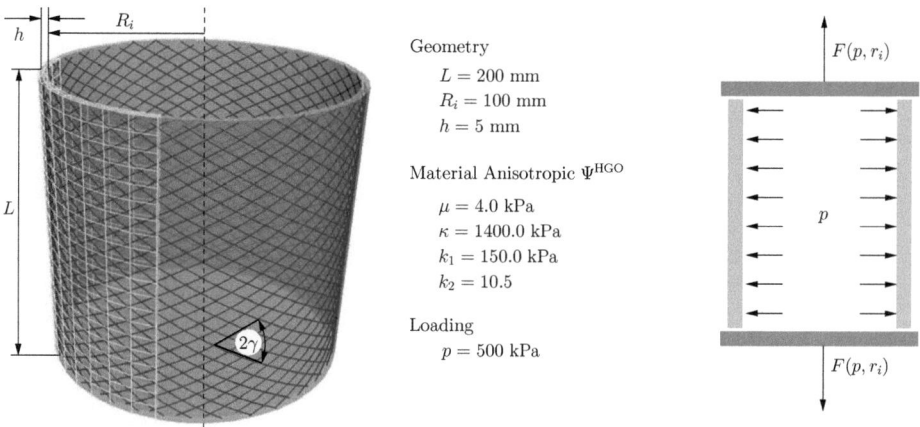

Fig. 4.8: Inflation of fiber-reinforced rubber tube. The fiber reinforcement and the discretized structure are sketched on the left and the loading condition on the right.

The fiber-reinforced rubber material is modeled with the Holzapfel-model Ψ^{HGO} with material parameters for the isotropic Neo-Hookean part and equal values for both fiber families, specified in Fig. 4.8. The fiber angle γ with respect to the circumferential direction is varied in the following computations to observe the typical *stretch inversion phenomenon* of such structural elements in the low pressure domain.

In Fig. 4.9 the resulting load-stretch curves are depicted. Above, the evolution of the circumferential stretch $\lambda_\theta = \frac{r}{R} = 1 + \frac{u_r}{R}$ is plotted versus the internal pressure, whereas below the evolution of the longitudinal stretch against the pressure is plotted, each for varying fiber angles γ from 30° to 40°. Note that for γ around 30° the tube initially *stretches longitudinally* but contracts circumferentially due to incompressibility of the material. Conversely, for γ above about 35° it initially *stretches circumferentially* but contracts longitudinally. These results agree well with the observations presented by Holzapfel and Gasser [139] and Nedjar [218]. The intriguing effect of the fiber-reinforcement successfully modeled by the anisotropic material law is clearly demonstrated.

4.6.4 Cyclic Inflation of a Viscoelastic Fiber-Reinforced Tube

We now study viscous effects in the fiber-reinforced rubber tube examined in the previous section. The structural problem, discretization and elastic material modeling are kept the same. However, we introduce viscoelasticity as discussed in Section 4.5, considering the behavior subject to cyclic inflation. We apply the internal pressure $p = p(t)p_0$ with $p_0 = 100$ kPa as

$$p(t) = \begin{cases} 0.125(1 - \cos(2\pi t)) & 0 < t \leq 0.5 \\ 0.125(1 - \cos(8\pi(t - 0.5))) + 0.25 & 0.5 < t \leq 2.0 \end{cases} \quad (4.126)$$

in a simulation with 400 timesteps and $\Delta t = 0.005$. The pressure load is plotted with respect to time in the inset of Fig. 4.10. We take one relaxation process into account, $m = 1$ and apply the parameters controlling viscoelasticity as $\tau^{\text{blk}} = 1.0$, $\beta^{\text{blk}} = 2.0$ for the isotropic part and $\tau^{\text{fib1,2}} = 1.0$, $\beta^{\text{fib1,2}} = 2.0$ for both reinforcing fiber families with an alignment angle $\gamma = 40°$. This simulation is taken as 'base result'.

The plots in Fig. 4.10 representing the circumferential stretch versus the internal pressure demonstrate the viscoelastic effect characterized by the hysteresis during cyclic inflation. The base result (red dash-dot) shows an increasing stretch due to viscous damping. The same increasing damping is present for a simulation with an alignment angle of $\gamma = 30°$ (green solid), however the stretch inversion phenomenon is still apparent in the low pressure regime. To examine the viscoelastic influence of the bulk contribution and the anisotropic fiber contribution we altered the corresponding parameters. The simulation with only the anisotropic fiber-related contribution is activated, $\beta^{\text{blk}} = 0.0$, $\beta^{\text{fib1,2}} = 2.0$ (black dashed), is hardly different from the original 'base result'. Only in the initial low pressure regime the lower stiffness due to the missing viscoelastic contribution is noticeable. However, the simulation with a viscoelastic element only for the isotropic bulk contribution, $\beta^{\text{blk}} = 8.0$, $\beta^{\text{fib1,2}} = 0.0$ results in a significantly different curve (blue solid). Due to the missing viscoelastic fiber contribution the stretch is generally larger. But the viscosity effect is reduced and only a small hysteresis is observable.

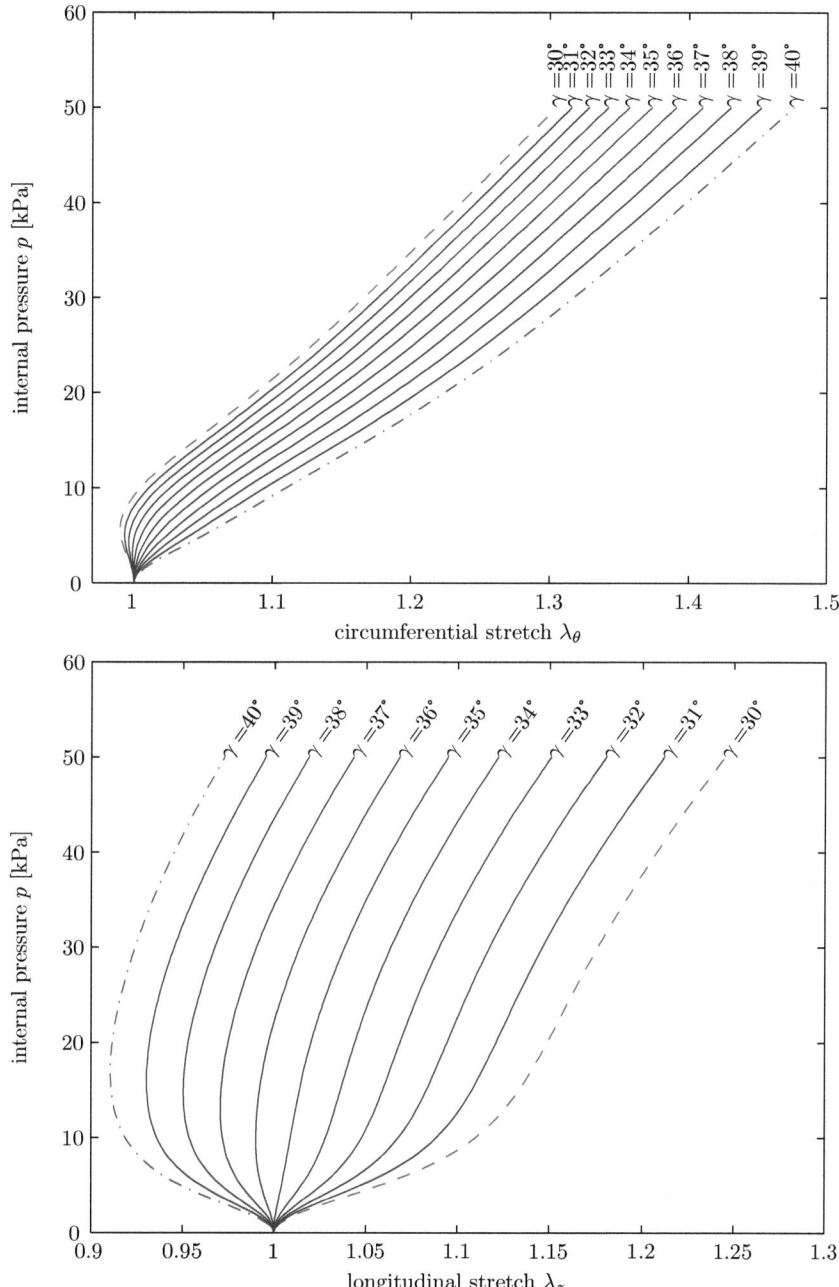

Fig. 4.9: Pressure-stretch results in circumferential (top) and longitudinal (bottom) direction for different fiber angles demonstrating the *stretch inversion phenomenon*.

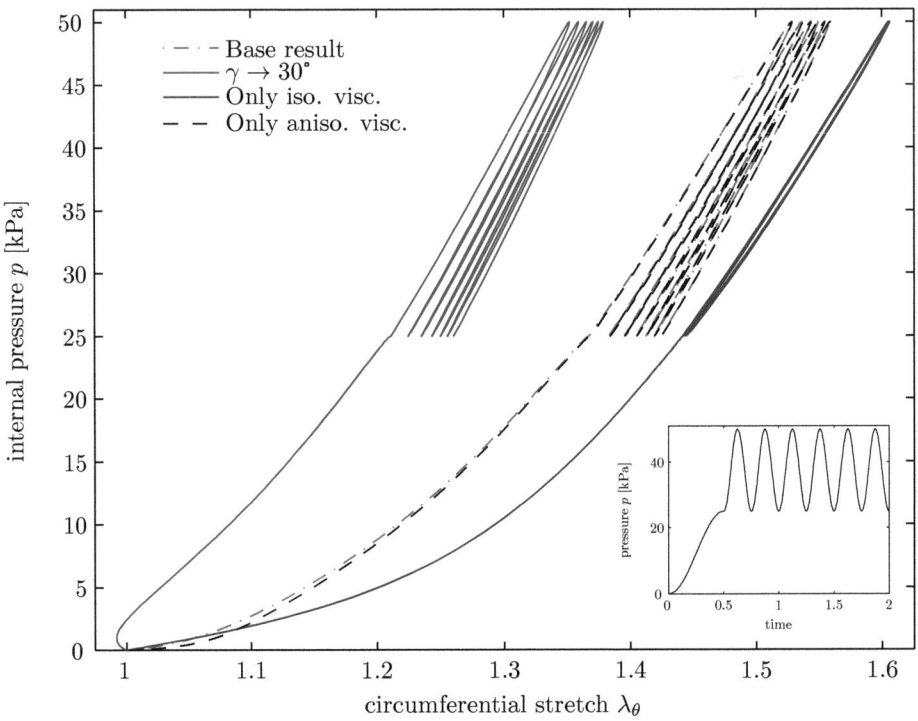

Fig. 4.10: Cyclic inflation of viscoelastic fiber-reinforced tube. Loading history (inset) and circumferential pressure-stretch results.

5. The Biomechanical Phenomenon of Remodeling

Remodeling can be defined as an evolution of microstructure, in our case biological tissue, by adapting to its environment. The important role of the mechanical environment in controlling remodeling is attested by several observations, for instance due to altered loads ranging from exercise to microgravity, due to interventional procedures such as stenting or bypass surgery, and due to developing diseases ranging from hypertension to atherosclerosis or the formation of aneurysms. There is a close connection with the phenomenon of growth which warrants a clear distinction of terms and related phenomena. Generally, there has been an increased realization that mathematical modeling plays a central role to better understand and predict these phenomena. Within this chapter we present an overview about mathematical and algorithmic treatment with emphasis on application for patient-specific cardiovascular computations. We start with a determination of terms and phenomena and shortly review current literature. Subsequently, we report the governing equations, describe three major approaches found in the literature and discuss shortcomings and improvements. Concluding this chapter we demonstrate numerical examples of idealized arterial sections, tendons, and bifurcation geometries.

5.1 Literature Review and Definition of Terms

Growth and remodeling processes take place in many biomechanical structures, ranging from bone (see the initiating work of Cowin [61, 58] for instance) to cartilage, to tendons, to arterial tissue, to name just a few, and many of these attracted distinct research attention. Whereas for the case of bone the high stiffness allows a decription in the small deformation regime, soft tissue premises a nonlinear, large deformation theory. Moreover, having this large range of examples in mind many more aspects of distinction emerge. Where a muscle adapting to load is directly connected with an increase in volume, increasing bone stiffness is related to a gain in density. We follow the classification of Taber [288] who defines the distinct processes of *morphogenesis, growth and remodeling*. We want to concentrate on growth and remodeling and thereby render growth more precisely as addition or depletion of mass, whereas remodeling is defined as *microstructural change within the biological structure at constant mass*. This is in accordance with Garikipati *et al.* [107, 109], Arruda *et al.* [11], Hariton *et al.* [121], and Guillou and Ogden [118], among others. Our treatment of remodeling in the following sections is confined to this latter definition. However, as growth and remodeling occur simultaneously and in a coupled fashion in biological tissues such as the artery, important contributions also in the field of growth or without this precise distinction are included in the following literature review.

Continuum level mathematical models of growth and remodeling in soft tissue have became an active area of research in recent years. One of the first attempts was that of Skalag et al. [276] in 1992. Subsequently, a formal three-dimensional framework for mechanically modulated volumetric growth of soft tissue has been published 1994 by Rodriguez et al. [252]. Herein, the modeling of overall growth deformation as a composition of deformation gradient mappings has been substantiated. They suggested to describe the overall growth deformation, denoted as $\boldsymbol{F}_{eg}$, as

$$\boldsymbol{F}_{eg} = \boldsymbol{F}_e \boldsymbol{G}, \tag{5.1}$$

where $\boldsymbol{G}$ is a symmetric tensor representing the growth deformation gradient and $\boldsymbol{F}_e$ represents the elastic deformation necessary to maintain overall compatibility.

This multiplicative decomposition into growth and deformation has been followed by the majority of researchers, for instance Taber [288], Rachev et al. [240], Cowin [60], Ambrosi and Mollica [4], Lubarda and Hoger [200, 199], Epstein and Maugin [80]. The latter have provided also theoretical considerations on the underlying framework of *open systems* (see also Holzapfel [137], Kuhl and Steinmann [187], and Garikipati et al. [107], among others).

However, a complete, continuum thermodynamics treatment of a system that is open with respect to mass to allow species concentration changes due to mass transport, such as nutrients transported within a fluid phase, is highly complex and exceeds today's simulation capabilities. In the case of the cardiovascular system, where normal arterial development as well as functional adaptations, responses to injury, and many disease processes occur by a cell mediated turnover of individual wall constituents at different rates, different extents, and in different biomechanical states, this is obviously true. It is because of the diverse repertoire of mechanosensitive cellular activities (such as migration, proliferation, apoptosis, synthesis and degradation of matrix, and production of adhesion molecules, vasoactive molecules, growth factors, matrix metalloproteinases and cytokines) that a complete coverage of all phenomena is not possible.

Thereby motivated, a promising approach was taken by Humphrey and colleagues [155] called the *constrained mixture* model. They suggest exploiting the full mixture equations for mass balance, hence modeling the production and removal of individual constituents, but enforce only a single momentum balance for the mixture. This concept has been further elaborated within this group (for instance Humphrey and Rajagopal [155], Gleason et al. [114], Alford et al. [2]) and very recently it has been included into a coupled fluid-solid-growth framework for cardiovascular simulations by Figueroa et al. [85].

Further selected publications on cardiovascular applications of growth models are the work of Watton and colleagues [330, 329] and Kroon and Holzapfel [178, 179, 180, 182] on growth of aneurysms, and the papers of Ambrosi et al. [3] and Kuhl et al. [186] on arteries. Menzel combines both growth and remodeling in a general anisotropic growth approach [212, 213].

5. The Biomechanical Phenomenon of Remodeling

Focusing on remodeling of (arterial) tissue, according to the previous definition, as remodeling of the microstructure, or even more specific as reorientation of the collagen fiber constituents within the material, Driessen and colleagues [72, 74, 75, 73] probably initiated research in this direction. However, Garikipati [109] reported several shortcomings of their initial approach, discussed in detail in Section 5.3.1. Hariton et al. [121, 122] proposed another fiber remodeling procedure and successfully applied it to an idealized human carotid bifurcation geometry. Their model is described in detail in Section 5.3.2. An algorithmic improvement is contributed by Himpel et al. [135] in linearizing the reorientation algorithm of one single fiber direction. However, extension to more fiber families as present in the arterial wall seems much more complicated. Both the latter and the former approaches are based on a fiber reinforced anisotropic material formulation as described in Section 4.3. In contrast, Kuhl and colleagues [184, 185] employ a continuum molecule chain material (cf. Section 4.4) in their remodeling approach, which is discussed in Section 5.3.3. Prior to a detailed review of each of these approaches, the governing equations of continuum mechanical remodeling are briefly reported.

5.2 Governing Equations

In this section we describe the mathematical description of remodeling, treated as a motion in material space or a configurational change. We thereby follow closely the work of Garikipati et al. [108, 109] and Arruda [11]. The kinematics of remodeling are assumed as illustrated in Fig. 5.1.

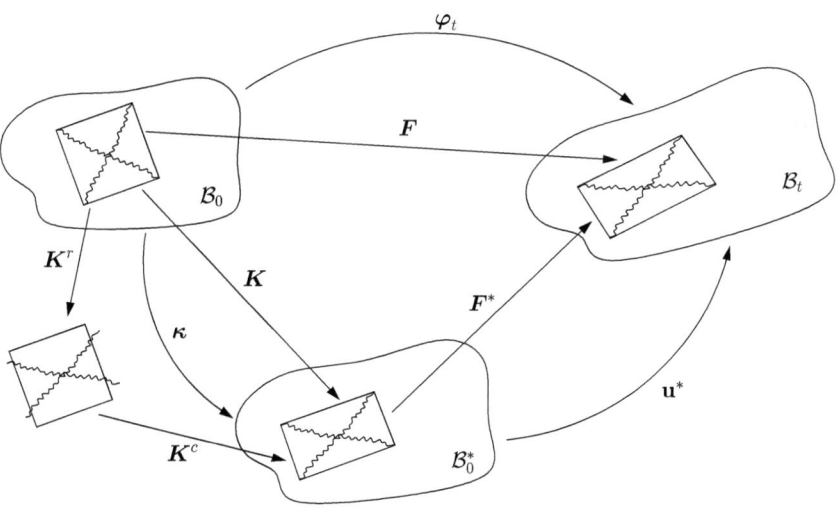

Fig. 5.1: The kinematics of remodeling, adapted from Garikipati et al. [108].

The overall deformation gradient $\boldsymbol{F}$ is established by a multiplicative decomposition

$$\boldsymbol{F} = \boldsymbol{F}^* \boldsymbol{K}^c \boldsymbol{K}^r \tag{5.2}$$

with $\boldsymbol{K}^r$ as the tangent map to the preferred remodeled state that a region would attain if it were free to remodel according to a local driver. Such a driving force could be the cellular mechanotransductive responses or, in a more phenomenological approach, the principal strain or stress field. A corresponding evolution law yields $\boldsymbol{K}^r$. This map is in general incompatible to constraints placed on the region by surroundings and $\boldsymbol{K}^c$ represents the compatibility restoring tangent map to the remodeled configuration. Therefrom $\boldsymbol{F}^*$ maps to the current configuration.

The mechanical theory is examined by minimizing the potential energy functional in the remodeled configuration $\mathcal{B}_0^*$, with $(\cdot)^*$ referring to this configuration, given as

$$\Pi_{\text{rem}}(\mathbf{u}^*, \boldsymbol{\kappa}) = \int_{\mathcal{B}_0^*} \Psi^*(\boldsymbol{F}^*, \boldsymbol{K}^c, \mathbf{X}^*) \mathrm{d}V^* - \int_{\mathcal{B}_0^*} \mathbf{f}^* \cdot (\mathbf{u}^* + \boldsymbol{\kappa}) \mathrm{d}V^* - \int_{\Gamma_N^*} \hat{\mathbf{t}}^* \cdot (\mathbf{u}^* + \boldsymbol{\kappa}) \mathrm{d}A^* \tag{5.3}$$

where $\boldsymbol{\kappa} = \mathbf{X}^* - \mathbf{X}$ is the motion of a point in material space (configurational change), $\mathbf{u}^*$ is the displacement, $\Psi^*(\boldsymbol{F}^*, \boldsymbol{K}^c, \mathbf{X}^*)$ is the stored energy function and is assumed to depend on the compatibility restoring motion $\boldsymbol{K}^c$ in addition to $\boldsymbol{F}^*$, $\mathbf{f}^*$ is the body force and $\hat{\mathbf{t}}^*$ is the surface traction. Stationarity is assumed with respect to both displacements $\mathbf{u}^*$ and $\boldsymbol{\kappa}$. Variational calculus yields the Euler-Lagrange equations

$$\text{Div}^* \boldsymbol{P}^* + \mathbf{f}^* = \mathbf{0} \qquad \text{in } \mathcal{B}_0^* \tag{5.4}$$

$$-\text{Div}^* \left(\Psi^* \boldsymbol{I} - \boldsymbol{F}^{*\text{T}} \boldsymbol{P}^* + \frac{\partial \Psi^*}{\partial \boldsymbol{K}^c}(\boldsymbol{K}^c)^{\text{T}} \right) + \frac{\partial \Psi^*}{\partial \mathbf{X}^*} = \mathbf{0} \qquad \text{in } \mathcal{B}_0^* \tag{5.5}$$

$$\boldsymbol{P}^* \mathbf{N}^* = \hat{\mathbf{t}}^* \qquad \text{on } \Gamma_N^* \tag{5.6}$$

$$\left(\Psi^* \boldsymbol{I} - \boldsymbol{F}^{*\text{T}} \boldsymbol{P}^* + \frac{\partial \Psi^*}{\partial \boldsymbol{K}^c}(\boldsymbol{K}^c)^{\text{T}} \right) \mathbf{N}^* = \mathbf{0} \qquad \text{on } \Gamma_N^*, \tag{5.7}$$

where we refer to Garikipati et al. [109] for a detailed derivation. The term $\frac{\partial \Psi^*}{\partial \boldsymbol{K}^c} \boldsymbol{K}^{c\text{T}}$ is a thermodynamic driving quantity arising from the change in stored energy Ψ^* corresponding to the change in configuration $\boldsymbol{K}^c$. It is identified to be stress-like and we refer to it as configurational stress $\boldsymbol{\Sigma} = \frac{\partial \Psi^*}{\partial \boldsymbol{K}^c} \boldsymbol{K}^{c\text{T}}$.

The remodeling is subjected to restrictions placed by the dissipation inequality for the mechanical theory (cf. Equation (2.49) in Section 2.3), written in terms of the Kirchhoff stress

$$\boldsymbol{\tau} = \det \boldsymbol{K} \frac{\partial \Psi^*}{\partial \boldsymbol{F}^*} \boldsymbol{F}^{*\text{T}}, \tag{5.8}$$

as

$$\boldsymbol{\tau} : (\dot{\boldsymbol{F}} \boldsymbol{F}^{-1}) - \frac{\partial}{\partial t}(\det \boldsymbol{K} \ \Psi^*) \geq 0, \tag{5.9}$$

with $\boldsymbol{K} = \boldsymbol{K}^c \boldsymbol{K}^r$.

This leads to the following reduced dissipation inequality, placing a restriction on the evolution law for $\boldsymbol{K}^r$ and on the functional form of the extra configurational stress $\boldsymbol{\Sigma}$ as

$$-\det \boldsymbol{K}(\boldsymbol{\mathcal{E}} + \boldsymbol{\Sigma}) : (\dot{\boldsymbol{K}}^c (\boldsymbol{K}^c)^{-1}) - \det \boldsymbol{K} \ \boldsymbol{\mathcal{E}} : (\boldsymbol{K}^c \dot{\boldsymbol{K}}^r \boldsymbol{K}^{-1}) - \det \boldsymbol{K} \frac{\partial \Psi^*}{\partial t} \geq 0 \tag{5.10}$$

where we introduced the Eshelby stress $\boldsymbol{\mathcal{E}} = \Psi^* \boldsymbol{I} - \boldsymbol{F}^{*\text{T}} \boldsymbol{P}^*$.

Remark This restriction due to dissipation states that in a purely mechanical theory dissipation would be positive for stiffening materials. This indicates that other thermodynamic phenomena, for instance of chemo-mechanical nature, need to be taken into account for consistency. It is widely accepted that a purely mechanical theory is thermodynamically inadmissible for remodeling processes stiffening the material (see also Menzel [212], Kuhl et al. [184], Himpel et al. [135]). Alternatively, exchange of energy and entropy among individual constituents could be considered, for instance in a mixture theory. However, according to Kuhl and Holzapfel [185], there is not yet a general agreement of how evolution laws for remodeling should be formulated. The evolution laws described in the following sections are all purely mechanically motivated, either stress or strain driven. The relevant driving force for remodeling is currently not clear. However, it has been shown for linear elasticity that the free energy attains an extremum if strain and stress share the same principal directions (see Cowin [59] or Vianello [317]).

In a nutshell, the remodeling approaches presented so far are merely an attempt to study the evolving tissue structure and allow a thermodynamically inconsistent analysis. These strategies cannot be related yet to the rigorous deformation response in a purely mechanical setting. However, they serve well to attain a more detailed and biomechanically senseful material configuration. Initial and boundary conditions play an important role, but are not yet well investigated. The same holds for any purely mechanically motivated evolution law. These limitations have to be considered when employing the approaches described below.

5.3 Fiber Remodeling and Review of Recent Approaches

5.3.1 Driessen's Approach

Driessen and colleagues [72, 74, 75, 73] have substantially contributed to the field of collagen fiber remodeling in cardiovascular tissue. In this section we chronologically describe their models, because many of the underlying ideas emerge in the following remodeling approaches as well. We also discuss some of the assumptions, implications and limitations inherent to their models.

Initially, the remodeling within a closed stented aortic heart valve was the application of the proposed remodeling approach of Driessen et al. [72, 74]. The key component is the introduction of a so-called *fiber orientation tensor* S_0 by scaling the structural tensor M_0 describing the anisotropic fiber in the undeformed configuration for every fiber family $i = 1, \ldots, N$ (see Section 4.3), with a probability distribution ϱ_i, yielding

$$A_0 := \sum_{i=1}^{N} \varrho_i M_{0i}, \tag{5.11}$$

where $\sum_{i=1}^{N} \varrho_i = 1.0$ holds, for instance $\varrho_i = 1/N$. The orientation tensor deforms in an affine manner such that the fiber orientation tensor S in deformed configuration is given as

$$A = \frac{F \cdot A_0 \cdot F^{\mathrm{T}}}{\Lambda^2} \tag{5.12}$$

where Λ^2 is the mean value of the square of the fiber stretch $\Lambda^2 \equiv \langle \lambda_{\text{fib}}^2 \rangle$ and $\langle \cdot \rangle$ denotes the average over the distribution space. The underlying constitutive law is presented as

$$\boldsymbol{\sigma}^{\text{fib}} = 3k(\boldsymbol{F}\boldsymbol{A}_0\boldsymbol{F}^{\text{T}} - \boldsymbol{Q}\boldsymbol{A}_0\boldsymbol{Q}^{\text{T}}) \qquad (5.13)$$

with $\boldsymbol{Q}$ being the rotation tensor of the deformation ($\boldsymbol{Q} = \boldsymbol{F}\boldsymbol{U}^{-1}$) and k the modulus of the fibers as a function of Λ^2.

As evolution of the orientation tensor they assume for the quasi-static case a first-order rate equation

$$\dot{\boldsymbol{S}} = \frac{1}{\tau_1}(\check{\boldsymbol{A}} - \boldsymbol{A}) \qquad (5.14)$$

with τ_1 being a time constant and $\check{\boldsymbol{A}}$ the stimulus for fiber reorientation, defined as a function of the Finger tensor $\boldsymbol{b}$:

$$\check{\boldsymbol{A}} = \frac{\boldsymbol{b}^\nu}{\text{tr}[\boldsymbol{b}^\nu]}. \qquad (5.15)$$

The parameter ν controls the degree of alignment with $\nu > 1$ representing a pronounced alignment.

Additionally, they take a fiber volume fraction ϕ into account which scales linearly between contributions of the fibers and the matrix material, yielding a Cauchy stress of the incompressible material

$$\boldsymbol{\sigma} = -p\boldsymbol{I} + (1 - \phi)\boldsymbol{\sigma}^{\text{blk}} + \phi\boldsymbol{\sigma}^{\text{fib}}. \qquad (5.16)$$

This volume fraction or amount of fibres ϕ is assumed to additionally change during remodeling, depending linearly on Λ^2. The corresponding evolution is again modeled by a first-order rate equation as

$$\frac{d\phi}{dt} = \frac{1}{\tau_2}[\phi_{ss}(\Lambda^2) - \phi] \qquad (5.17)$$

with τ_2 as time constant and ϕ_{ss} as a steady-state fiber amount.

This methodology highlights the three typical ingredients of a remodeling approach: (1) a constitutive law depending on a varying fiber structure is required, here represented by (5.16); (2) an evolution equation has to be defined to describe the time-depending process and usually first-order rate equations are employed such as equations (5.14) and (5.17); (3) the remodeling driver has to be specified, in this case depending on the strain (5.15).

In their papers Driessen et al. [72, 74] have modeled the fiber distribution of a heart valve. However, as pointed out by Garikipati et al. [109], their strategy has two substantial shortcomings. One is related to the fiber orientation tensor which fails to distinguish between the material characteristics of cubic orthotropy and isotropy. The other is related to the evolution law which drives *any* initial fiber orientation tensor to the unit tensor $\boldsymbol{I}$ without having any deformation ($\boldsymbol{b} \equiv \boldsymbol{I}$). This does not satisfy the physiologically intuitive observation that fiber orientations remain unchanged without any deformation. Another complication of their model is the combination of reorientation and volume fraction adjustment of the fibers. This contradicts with the general idea that remodeling is only a process of changing microstructure, but separated from growth.

In a later contribution Driessen et al. [75] resolved the limitation of their initial proposal that resulted in fiber directions aligned with the principal strain directions. Furthermore, they introduced a preferred fiber direction situated in between the principal stretch directions to account for the fact that in blood vessels the principal strain directions are in general oriented axially and circumferentially, whereas the fiber orientation shows a helical arrangement with changing pitch. The degree of alignment of the preferred fiber direction with the principal stretch direction is controlled by another parameter. In this contribution they restricted their remodeling approach to changes only in fiber orientation, keeping fiber type, thickness, stiffness and especially content constant. By this approach the typical helical arrangement of collagen fibers in the arterial wall (see Rhodin [246]) has been obtained.

Recently, another adjustment of the remodeling procedure is proposed by Driessen et al. [73]. The major change was the reintroduction of the angular distribution of collagen fibers, required to accurately describe the complex biaxial mechanical behavior of the aortic valve. This introduces a further parameter, the dispersity of the fibers, in addition to the preferred fiber direction. This approach is based on the formulation of Gasser et al. [110] who introduce a *fiber dispersion* into the anisotropic material formulation of Holzapfel [140] (see Section 4.3.3). However, it has been recently shown by Federico and Herzog [81] that at least for an analytical usage this formulation is restricted to fibers in tension and with weak dispersion around a main direction. Although this is the case for blood vessels it reflects a limitation of this model.

Generally, Driessen's variant approaches can be characterized as consisting of several interrelated features. The combination of fiber orientation, distribution and/or dispersion complicates identification of influencing parameters and interpretation of the resulting structure. Another major question is the driving parameter, in most cases the strain. Although this question is not answered yet there are indications that the stress would be more appropriate as driving force (see Taber and Humphrey [290] and the discussion in Kuhl and Holzapfel [185]). Although Driessen and colleagues have proven the successfull application of their remodeling strategies for aortic valves and blood vessels we prefer different remodeling approaches, described in the following.

5.3.2 Hariton's Approach

Hariton et al. [121, 122] present a remodeling approach which is described in this section. Their major point is the choice of *stress* as driving force for fiber remodeling. They argue that the artery remodels to restore circumferential wall stresses due to pressurization and wall shear stress due to blood flow to 'normal' levels (see also Taber and Humphrey [290]). Restricting their model to a varying collagen fiber angle while keeping collagen content, type and thickness constant, they propose that local principle tensile stress magnitude and direction modulate the fiber alignment in arteries.

We review their model which has the popular anisotropic constitutive law of Holzapfel [140], described in Section 4.3.1, as its basis. The Cauchy stress at a material point is given by

$$\boldsymbol{\sigma} = \lambda_i^\sigma \, \boldsymbol{\phi}_i^\sigma \otimes \boldsymbol{\phi}_i^\sigma \tag{5.18}$$

with the principal stresses $\lambda_i^\sigma, i = 1, 2, 3$ and their corresponding principal directions $\boldsymbol{\phi}_i^\sigma$ sorted such that $\lambda_1^\sigma \geq \lambda_2^\sigma \geq \lambda_3^\sigma$. They assume that the collagen fibers lie within a plane spanned by the two vectors $\boldsymbol{\phi}_1^\sigma$ and $\boldsymbol{\phi}_2^\sigma$ and symmetrically align relative to $\boldsymbol{\phi}_1^\sigma$. In terms of the angle of alignment γ the unit vectors along the two fiber families in the current configuration are defined as

$$\mathbf{m} = \cos\gamma \, \boldsymbol{\phi}_1^\sigma + \sin\gamma \, \boldsymbol{\phi}_2^\sigma, \qquad \mathbf{n} = \cos\gamma \, \boldsymbol{\phi}_1^\sigma - \sin\gamma \, \boldsymbol{\phi}_2^\sigma. \tag{5.19}$$

The key component with respect to their hypothesis of stress as remodeling driver is the definition of the so-called *modulation function* which they define as

$$\tan\gamma = \mathcal{M}\left(\frac{\lambda_2^\sigma}{\lambda_1^\sigma}\right). \tag{5.20}$$

In their presented work they employ the simplifying assumption

$$\mathcal{M}\left(\frac{\lambda_2^\sigma}{\lambda_1^\sigma}\right) = \frac{\lambda_2^\sigma}{\lambda_1^\sigma}. \tag{5.21}$$

Future work could introduce additional, possibly nonmechanical factors into this modulation function. Interpretation of Equation (5.21) reveals that the fibers would perfectly align with the direction of uniform tension, whereas in the case of equal biaxial principal stresses $\lambda_1^\sigma = \lambda_2^\sigma$ the fibers align exactly in between the two directions $\boldsymbol{\phi}_1^\sigma$ and $\boldsymbol{\phi}_2^\sigma$. It is remarked that the alignment of the fibers is defined in the deformed configuration and thus the unit vectors in the reference configurations are obtained by a pull-back operation:

$$\mathbf{m}_0 = \frac{\boldsymbol{F}^{-1}\mathbf{m}}{|\boldsymbol{F}^{-1}\mathbf{m}|} \quad \text{and} \quad \mathbf{n}_0 = \frac{\boldsymbol{F}^{-1}\mathbf{n}}{|\boldsymbol{F}^{-1}\mathbf{n}|} \tag{5.22}$$

Hariton's proposed model rests upon the assumption that the collagen fibers remodel into a mechanically optimal configuration. However, the actual evolution process is not explicitly specified. They employ an iterative procedure starting usually with a random initial fiber configuration. In each step the principal stresses and corresponding fiber directions are evaluated until the point where the stress fields determined in two consecutive computations are sufficiently close.

Thus, they renounce to specify an evolution equation. On the one hand this seems to be a plausible assumption, because a thermodynamically and biomechanically correct model is not yet available. On the other hand for a truely time-dependent simulation an evolution equation is necessary. Moreover, an evolution law assuming a diminishing remodeling capacity would decrease oscillations in fiber directions. Furthermore, the specification of a termination criterion would then be dispensable. We discuss such advancements of the present remodeling approach in Section 5.4.

An important remark emphasized by Hariton *et al.* states that "due to the highly nonlinear nature of the problem it is not clear whether there is an optimal configuration of the fibers or, in the case that such a configuration exists, if this configuration is unique" ([121], p. 166). We confirm their observation that usually the proposed approach converges to almost identical results with different initial conditions. However, our performed numerical examples (see Section 5.5) also demonstrate that the solution of such problems is computationally demanding and sometimes not very robust. Certainly, further research is necessary in this respect.

5.3.3 Kuhl's Approach

A very recent remodeling approach is proposed by Kuhl and Holzapfel [185]. It is based on the continuum molecule chain material model, described in detail in Section 4.4. As discussed, one of the crucial benefit of this material model is the fact that the underlying parameters are physically motivated. Admittedly, the application and experimental parameter fitting of this model for arterial walls is not available in literature and we therefore rely on the parameters reported in the qualitative studies of Kuhl and Holzapfel [185].

We recall that the continuum molecule chain material model is capable of representing isotropy, transverse anisotropy and orthotropy by defining different aspect ratios to the unit-cell dimensions. This option is utilized by the proposed remodeling approach. Whereas in a preliminary paper by Kuhl *et al.* [184] the concept was to rotate a predefined unit-cell for remodeling of transversely isotropic tissue such as tendons, the present approach directly changes the unit-cell dimensions following a specified remodeling strategy. Therefore a smooth transition between isotropy and anisotropy is enabled. Additionally, the parameterization of unit-cell dimensions as remodeling variables is advantageous with respect to numerical handling in contrast to a parameterization based on angles and trigonometric functions.

The underlying assumption of the remodeling process in form of a microstructural rearrangement is to allow the fiber direction to rotate in response to the current mechanical stress response. While this is intuitively modeled in the previous approach by allowing the fiber angle to change, it corresponds to a change in alignment *and* dimensions of the fiber unit-cell in the approach at hand. Therefore, the Cauchy-stress tensor as remodeling driving force is decomposed into its principal values and directions as in (5.18), repeated here as

$$\boldsymbol{\sigma} = \lambda_i^\sigma \, \boldsymbol{\phi}_i^\sigma \otimes \boldsymbol{\phi}_i^\sigma. \tag{5.23}$$

The two fundamental hypotheses of the remodeling process are:

1. The characteristic material axes $\mathbf{a}_i$ of the unit-cell instantaneously align with the principal directions of the Cauchy-stress $\boldsymbol{\phi}_i^\sigma$.

2. The unit cell dimensions a_i adapt gradually with respect to the positive eigenvalues $\lambda_i^{\sigma+}$.

Concerning the first postulate, Kuhl and Holzapfel remark that the general idea of network models is that the unit cell is taken to deform in principal stretch space (see also Boyce and

Arruda [46] and Bischoff et al. [32]). We point out that changing the material axes of the unit cell is inherently related to a change in the stress response and therefore goes along with a change in internal energy of the underlying structure. The necessary additional driving energy and entropy are not taken into account following a purely mechanically motivated remodeling strategy as discussed in Section 5.2.

The second postulate resembles the idea that collagen fibers in cardiovascular tissues are located between the two maximal principal stress directions which is reflected in a corresponding unit-cell with specific dimensions defined as

$$a_i = r_0 \frac{\lambda_i^{\sigma+}}{\|\lambda_i^{\sigma+}\|}. \tag{5.24}$$

For some exemplary principal stress relations the resulting unit-cells are illustrated in Fig. 5.2. The assumption that only positive stresses, i.e. tension forces drive the remodeling is reflected in the definition of $\lambda_i^{\sigma+} = \lambda_i^\sigma$ for $\lambda_i^\sigma > 0$ and $\lambda_i^{\sigma+} = 0$ for $\lambda_i^\sigma \leq 0$.

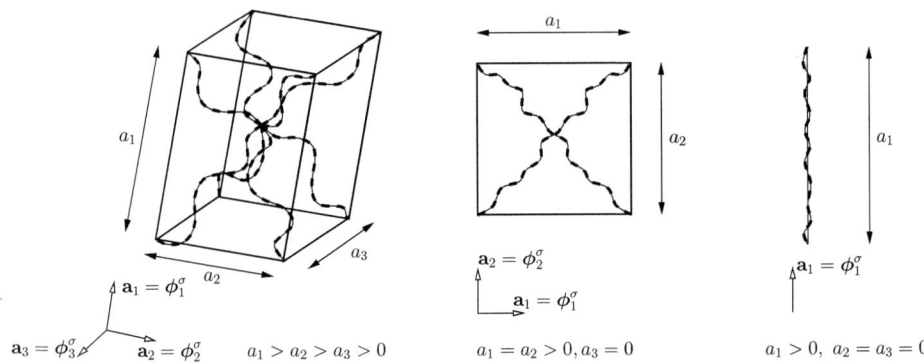

Fig. 5.2: Different unit-cells illustrating the remodeling cases of orthotropy (left), transverse isotropy with two orthogonal fiber directions (middle), and transverse isotropy with one fiber direction (right), together with the corresponding principal stress relations.

In contrast to the merely iterative remodeling process proposed in the previous approach, Kuhl and Holzapfel explicitly specify a remodeling evolution law. The remodeling of the microstructural unit-cell dimensions is assumed to obey

$$\frac{\mathrm{d}}{\mathrm{d}t} a_i = \tau_\mathrm{rem} \left(\frac{\lambda_i^{\sigma+}}{\|\lambda_i^{\sigma+}\|} - \frac{a_i^0}{r_0} \right) \exp(-\tau_\mathrm{rem} t) r_0 \tag{5.25}$$

with τ_rem as remodeling relaxation parameter and a_i^0 as unit-cell dimensions in the initial undeformed configuration. An alternative formulation is given by a time integration yielding the explicit update for the cell dimensions

$$a_i = \left(\frac{\lambda_i^{\sigma+}}{\|\lambda_i^{\sigma+}\|} - \frac{a_i^0}{r_0} \right) (1 - \exp(-\tau_\mathrm{rem} t)) r_0 + a_i^0 \tag{5.26}$$

5. The Biomechanical Phenomenon of Remodeling

The remodeling approach based on the microstructural chain network unit-cell can be summarized as

$$\mathbf{a}_i^0(t) \equiv \boldsymbol{\phi}_i^\sigma(t) \quad \text{and} \quad a_i(t) \rightsquigarrow r_0 \frac{\lambda_i^{\sigma+}(t)}{\|\lambda_i^{\sigma+}(t)\|}. \tag{5.27}$$

In Section 5.5 we present numerical examples for this remodeling approach.

5.4 Suggested Advancements of Hariton's and Kuhl's Remodeling Approaches

In this section we discuss strengths and weaknesses of the presented approaches and propose some ideas how they might be advanced. Both approaches certainly result in reasonable fiber patterns and are in general applicable to successfully recover physiological fiber alignments (as demonstrated below in Section 5.5 which provides numerical examples). However, our comparisons between Kuhl's and Hariton's approach revealed some potential points of discussion.

Regarding Hariton's approach, the apparent missing ingredient is the definition of an evolution law. Both Driessen and Kuhl assume a decreasing remodeling capacity and therefore a determined fiber state is finally reached in their strategies. In contrast, Hariton et al. [122] report that they apply their remodeling process until the stress states of two consecutive analyses are sufficiently similar which is somewhat arbitrary. Certainly, the evolution equations of both Driessen and Kuhl are mere methodological assumptions and seem to have no physiological or biomechanical motivation. We nevertheless suggest taking into account a certain time-dependence for the *modulation function* $\mathcal{M}$ representing a decreasing remodeling capacity. This would allow a true time-dependent simulation and opens the door for temporal multiscale simulations.

Considering Kuhl's remodeling approach, a major appealing characteristic is the underlying material law based on a continuum chain network theory. The benefit is the physical interpretability of the material parameters. Further, with the unit cell dimensions as parameters for remodeling, no angular variables are involved which are cumbersome to linearize and typically result in difficulties with respect to objectivity in large deformations. Himpel et al. [135] presented a consistent linearization strategy for angular remodeling of *one* fiber direction. However, while this formulation already has significant complexity, the extension to more than one fiber family seems to be even more delicate. In contrast, the scalar valued parameters of the unit-cell involved in Kuhl's approach are much easier to handle. Interestingly, the same experience in the research of shell finite elements lead to the widely accepted finding that in large deformations the displacements (or the displacements of the shell directors) are preferred over rotations as unknowns.

One decisive parametric connection between the unit cell and the chain is the initial end-to-end length r_0 (see Section 4.4). Kuhl and Holzapfel claim that during remodeling driven by their evolution (Equation (5.25) or (5.26)) this initial chain length $r_0 = \sqrt{a_i^2}$ stays constant.

This claim is correct in a remodeling case where just one relevant cell dimension evolves and both other dimensions are reduced to zero. In that case, for $t \to \infty$, one cell dimension, for instance a_1, reaches r_0 and both a_2 and a_3 tend to zero. We repeat Equation (5.26) for this case:

$$a_1 = \left(\underbrace{\frac{\lambda_1^{\sigma+}}{\|\lambda_1^{\sigma+}\|}}_{:=1} - \frac{a_1^0}{r_0}\right)(1-\exp(-\tau_{\text{rem}}t))r_0 + a_1^0 \xrightarrow{t\to\infty} a_1 = r_0 \qquad (5.28)$$

$$a_{2,3} = \left(\underbrace{\frac{\lambda_{2,3}^{\sigma+}}{\|\lambda_{2,3}^{\sigma+}\|}}_{:=0} - \frac{a_2^0}{r_0}\right)(1-\exp(-\tau_{\text{rem}}t))r_0 + a_{2,3}^0 \xrightarrow{t\to\infty} a_{2,3} = 0. \qquad (5.29)$$

However, we would like to emphasize that this claim is incorrect in the general case. For example, in the case where two cell dimensions emerge, for instance a_1 and a_2, representing transverse isotropy with two orthogonal fiber directions, the situation differs:

$$a_{1,2} = \left(\underbrace{\frac{\lambda_{1,2}^{\sigma+}}{\|\lambda_{1,2}^{\sigma+}\|}}_{:=1} - \frac{a_2^0}{r_0}\right)(1-\exp(-\tau_{\text{rem}}t))r_0 + a_{1,2}^0 \xrightarrow{t\to\infty} a_{1,2} = r_0 \qquad (5.30)$$

$$a_3 = \left(\underbrace{\frac{\lambda_3^{\sigma+}}{\|\lambda_3^{\sigma+}\|}}_{:=0} - \frac{a_3^0}{r_0}\right)(1-\exp(-\tau_{\text{rem}}t))r_0 + a_3^0 \xrightarrow{t\to\infty} a_3 = 0. \qquad (5.31)$$

The resulting initial chain length equals $r_0\sqrt{2}$. For intermediate configurations, r_0 does not stay constant with this formulation, either. By implication, not only the directed stiffness due to the fiber alignment of the underlying material model changes during remodeling, but also the 'basis stiffness', due to alterations in the chain stiffness itself. This renders an identification of r_0 as material parameter impossible.

To circumvent this issue we suggest to introduce a rescaling of the cell dimensions after each remodeling step, such that

$$\tilde{a}_i = \left(\frac{\lambda_i^{\sigma+}}{\|\lambda_i^{\sigma+}\|} - \frac{a_i^0}{r_0}\right)(1-\exp(-\tau_{\text{rem}}t))r_0 + a_i^0, \qquad (5.32)$$

or

$$\frac{\mathrm{d}}{\mathrm{d}t}\tilde{a}_i = \tau_{\text{rem}}\left(\frac{\lambda_i^{\sigma+}}{\|\lambda_i^{\sigma+}\|} - \frac{a_i^0}{r_0}\right)\exp(-\tau_{\text{rem}}t)r_0, \qquad (5.33)$$

and $a_i = \dfrac{r_0}{\sqrt{\tilde{a}_i^2}}. \qquad (5.34)$

Thus the unit-cell always emerges in a way that the initial chain length stays constant.

Another issue with Kuhl's evolution equation can be observed, especially for the example of the idealized artery (see Section 5.5.4). Their remodeling driver is defined according to the hypothesis that for positive eigenvalues (tension), the corresponding cell length grows, but for negative eigenvalues it decreases. That implies that compression is needed to decrease a cell dimension.

The stress state in a tube under axial stretch and internal pressure is however typically characterized by tension in both the circumferential and axial direction, and compression in the radial direction. Only in a very limited region the incompressibility of the material introduces compression in axial or circumferential direction. This is illustrated in the sketch of a section through the axial-radial plane of a thick-walled ring in Fig. 5.3. The deformed state is depicted for an internal pressure loading with a free edge at the top. Due to the incompressibility of the material, a bulge emerges. However, if the top edge would be vertically fixed, the region of the bulge would sustain pressure, whereas the rest of the ring sustains tension.

Moreover, it is evident that the magnitudes of the two tension states differ. This should be reflected in the remodeling of the fibers by aligning more into the high tension direction and less into the lower tension direction. However, such a remodeling process is not possible with Kuhl's evolution law, as only the sign of the eigenvalues but not the quantity control the increase or decrease of the corresponding unit-cell dimension.

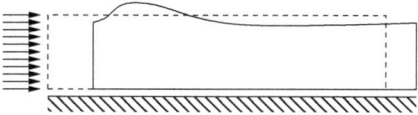

Fig. 5.3: Sketch of a axial-radial section of an incompressible ring under internal pressure. The deformed state is depicted for a free upper edge.

We propose to advance their law by relating the evolution of one cell dimension to the quotient of the corresponding eigenvalue λ_i^σ and the maximal eigenvalue λ_{max}^σ. This yields the explicit update for the cell-dimensions (cf. (5.26)):

$$a_i = \left(\left(\frac{\lambda_i^\sigma}{\lambda_{max}^\sigma} \right)^+ - \frac{a_i^0}{r_0} \right) (1 - \exp(-\tau_{rem} t)) r_0 + a_i^0 \qquad (5.35)$$

with the understanding that $(\cdot)^+$ corresponds to only positive values. With such a rule the dimension corresponding to the maximal stress eigenvalue λ_{max}^σ finally reaches the maximum of r_0, whereas the dimension corresponding to the second largest positive eigenvalue reaches a value defined by the ratio $\frac{\lambda^\sigma}{\lambda_{max}^\sigma} r_0$. By taking the rescaling, described in the previous paragraph, into account, the initial cell dimension r_0 is always conserved.

If a rate equation according to (5.25) is preferred, the evolution is suggested to take the form

$$\frac{d}{dt} a_i = \tau_{rem} \left(\left(\frac{\lambda_i^\sigma}{\lambda_{max}^\sigma} \right)^+ - \frac{a_i^0}{r_0} \right) \exp(-\tau_{rem} t) r_0, \qquad (5.36)$$

preserving the same benefits. With this modifications of the remodeling evolution we obtain reasonable and qualitatively physiological fiber patterns of arteries, as demonstrated in detail in Section 5.5.4.

Regarding Hariton's approach, the apparent missing ingredient is the definition of an evolution law. Both Driessen and Kuhl assume a decreasing remodeling capacity and therefore a determined fiber state is finally reached in their strategies. In contrast, Hariton et al. [122] report that they apply their remodeling process until the stress states of two consecutive analyses are sufficiently similar which is somewhat arbitrary. Certainly, the evolution equations of both Driessen and Kuhl are mere methodological assumptions and seem to have no physiological or biomechanical motivation. We nevertheless suggest taking into account a certain time-dependence for the *modulation function* $\mathcal{M}$ representing a decreasing remodeling capacity. This would allow a true time-dependent simulation and opens the door for temporal multiscale simulations. Finally, another option for advancing Hariton's approach could involve including additional mechanical factors into the remodeling capacity, for instance in a type of 'inverse damage' formulation.

5.5 Numerical Examples

5.5.1 Idealized Human Carotid Artery Model

As a first numerical example we present an idealized section of a human artery modelled as straight thick-walled cylinder. We chose geometric dimensions, material properties, loading, and boundary conditions summarized in Table 5.1 following Hariton et al. [121] which represents a human common carotid artery model, see also Delfino et al. [64].

The finite element model of a quarter of the carotid consists of twelve elements in circumferential direction, ten elements in axial directions and ten elements along the thickness. Because of the high slenderness of every element we apply the `Sosh8` solid-shell element formulation described in Section 3.5.2. The nonlinear problem is solved in a quasi-static solution procedure gradually applying the loading within ten equal load steps. At every load-step a remodeling step is conducted following the approach of Hariton et al. described in Section 5.3.2.

The result of this example is illustrated in Fig. 5.4. The color scale represents the first principal Cauchy-stress value which gradually decreases from about 180 kPa at the inside to 30 kPa at the outside of the artery. The stress distribution is in very good accordance to the reported distribution of Hariton et al. [121] The fiber alignment is illustrated by line segments at every node. Note that at some isolated nodes the line segments look cluttered due to postprocessing issues. Overall, it is evident that the fiber angle with respect to the circumferential direction gradually rises from inside to outside. To further clarify the resulting fiber pattern it is illustrated separately for the inner, middle and outer layer in a picture series in Fig. 5.5.

Table 5.1: Parameters of idealized artery model.

	Parameter	Value
Geometry	Inner Radius	3.1 mm
	Thickness	0.9 mm
	Length	4.0 mm
Material	μ	33.74 kPa
	k_1	13.9 kPa
	k_2	13.2 [-]
Loading	Internal Pressure	16 kPa
	Axial Stretch	10 %
Initial fiber angle	γ_0	39°

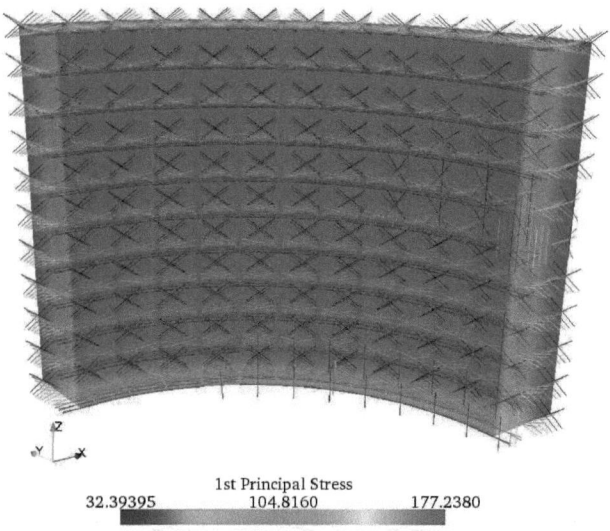

Fig. 5.4: Fiber alignment with first principal Cauchy-stress for idealized human artery model.

In Fig. 5.6 we plot the resulting fiber alignment angles through the thickness with varied bulk modulus κ, evaluated at one Gauss-point of each element. All curves agree well with the general understanding that the transmural pitch of the collagen fiber helix increases from the inner to the outer wall. For instance, Holzapfel et al. [145] found a mean angle of 5°, 7° and 49° for the intima, media and adventitia, respectively of a healthy human iliac artery, while Holzapfel et al. [141] reported 8.4° for the media and 41.9° for the adventitia of a human aorta.

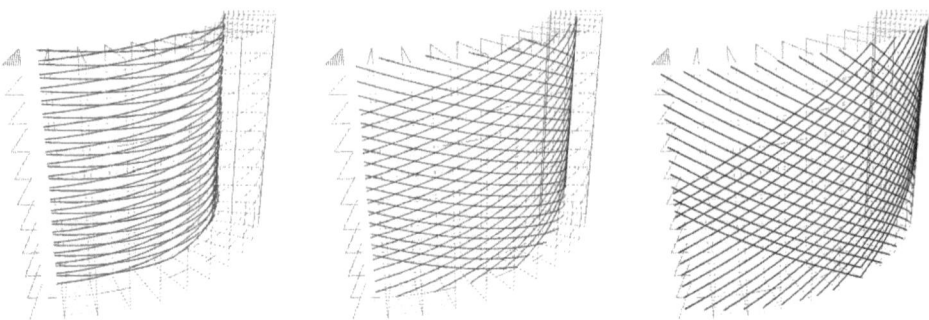

Fig. 5.5: Fiber pattern for the inner, middle and outer layer (from left to right) of the idealized human artery model.

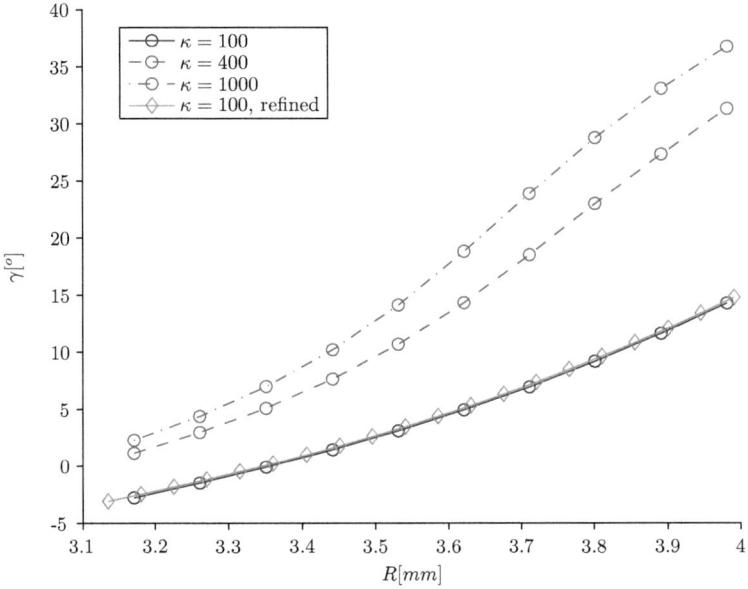

Fig. 5.6: Fiber alignment angle along the arterial wall thickness.

The resulting curve for $\kappa = 1000$ kPa agrees very well with the results presented by Hariton et al. [121], confirming the successful implementation of their remodeling approach. However, the other curves for a lower bulk modulus show a relatively strong dependence on the incompressibility constraint, enforced by the penalty method. Where the volume of the deformed artery for $\kappa = 1000$ kPa is within an acceptable tolerance (103% of the initial volume), it is significantly larger for $\kappa = 400$ kPa (107%) and beyond an acceptable tolerance for $\kappa = 100$ kPa (124%). The diamond marked curve depicts the result for a mesh refinement in thickness direction (20 elements) which confirms that already ten elements across the thickness result in a converged fiber pattern. This also agrees with observations by Hariton *et al.*

5.5.2 Idealized Human Carotid Bifurcation Model

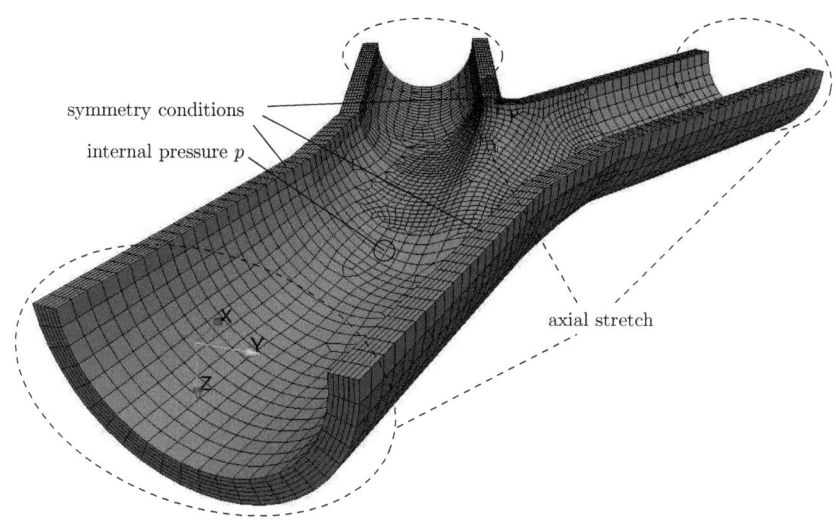

Fig. 5.7: Idealized carotid bifurcation model with mesh and boundary conditions.

To investigate remodeling in more complex geometrical configurations the second example considers remodeling of a bifurcation. We present an idealized carotid bifurcation model consisting of three cylinders of approximately physiological radii and a smoothed transition region to model the bifurcation. A similar example is discussed by Hariton *et al.* [122] who apply their remodeling approach to a geometric model of a human carotid bifurcation which has been investigated by Delfino *et al.* [64]. Therein a very detailed finite element model is presented with specific thickness and radius data for the stress-free and in vivo configurations.

In accordance to Hariton *et al.* we apply an internal pressure and an axial stretch to simulate in vivo longitudinal residual stretches (see Fig. 5.7). The material parameters are the same as in the previous example and we apply Sosh8 and EAS9 elements, respectively, depending on each element's slenderness.

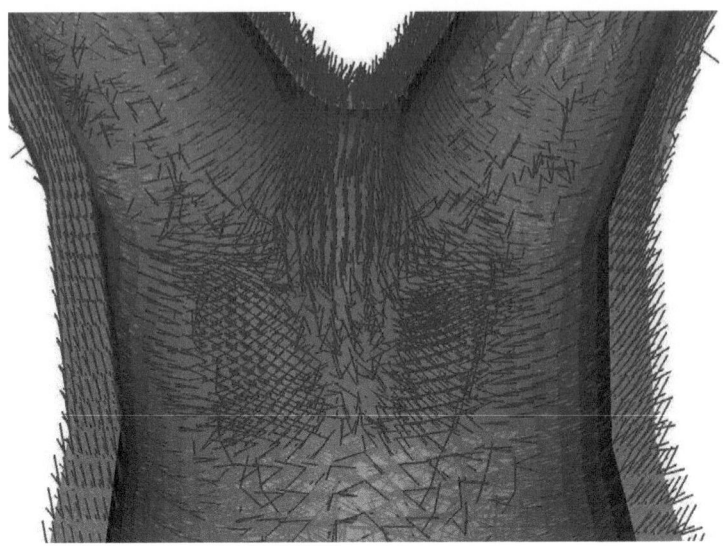

Fig. 5.8: Fiber alignment in idealized carotid bifurcation model as seen from inside.

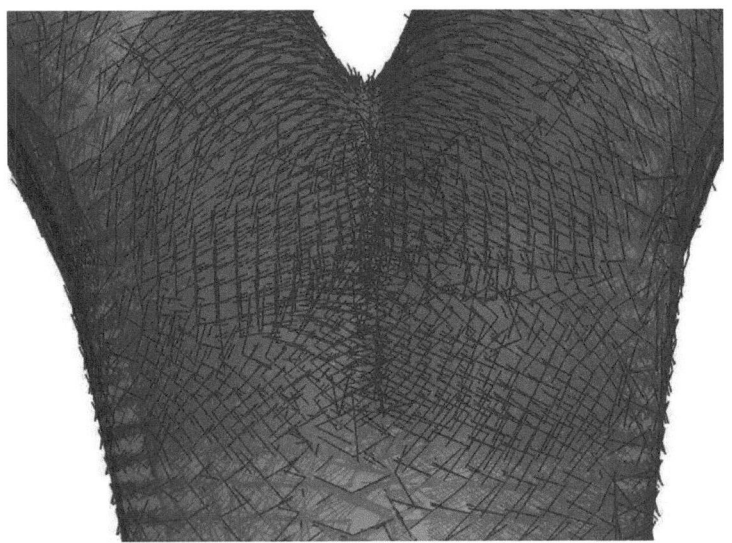

Fig. 5.9: Fiber alignment in idealized carotid bifurcation model as seen from outside.

As this geometric model is just a rough estimation with respect to geometry, residual stress and strain field we do not discuss the simulation results in detail but concentrate on the remodeling procedure. The resulting fiber configuration at the bifurcation (the apex) is illustrated in Fig. 5.8 from the inside and Fig. 5.9 from the outside. Focusing on the bifurcation region, the fiber morphology at the apical ridge can be identified to span tendon-like across the apex. Histological data reported by Finlay et al. [90] and Rowe et al. [253] confirms such a fiber structure. Moreover, these pictures are in perfect agreement with the remodeling results of Harriton et al. [122].

Note that the fiber pattern in the cylindrical regions away from the bifurcation is characterized by an alignment angle gradually increasing through the thickness from inside to outside. This resembles the previous example of a prestretched and pressurized cylindrical artery.

This example demonstrates that the remodeling strategy is capable to reproduce physiologically feasible fiber patterns also in more complex geometries. For even more complex geometries stemming from patient-specific CT-data we refer to Chapter 6.

5.5.3 Idealized Tendon Model

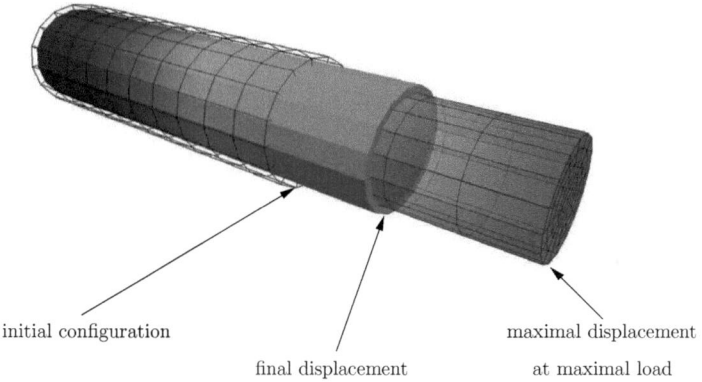

Fig. 5.10: Tendon displacement during the remodeling process.

Another example considers Kuhl's remodeling approach discussed in Section 5.3.3. This time not an arterial wall is modeled, but an idealized human tendon. The example is taken from Kuhl and Holzapfel [185] to confirm a correct implementation and illustrate their approach. Therefore, the geometry and the applied material parameters of the continuum chain network model described in Section 4.4 are also taken from this publication and summarized in Table 5.2. Note that we chose mm as unit length which together with Boltzmann's constant and the temperature sets the energy unit to be much higher than in the previous examples. The applied force, pressure and stretch are scaled accordingly. As no physiological parameters are available for this material law we stick to this rather qualitative study.

Table 5.2: Parameters of cylindrical tendon.

	Parameter	Value
Geometry	length	10.0 mm
	area	10.0 mm^2
Material	λ	27.293 GPa
	μ	3.103 GPa
	k	1.381×10^{-23} JK^{-1}
	θ	310 K
	N	7×10^{21} m^{-3}
	L	1.594 [-]
	A	1.365 [-]
	r_0	1.0 [-]
	τ_{rem}	0.025 [-]
Loading	F	300 kN
Initial unit-cell	isotropic	$r_0/\sqrt{3}$

The tendon is modeled by 640 EAS9-elements. It is fixed in z-direction on one side and in all directions at the middle node of this side. The loading is applied as a single force on the middle node of the other side and constraints are used to keep the load surface flat. The load-curve is a ramp with 12 load-steps for the linear increase and 138 steps to arrive at the final remodeled state.

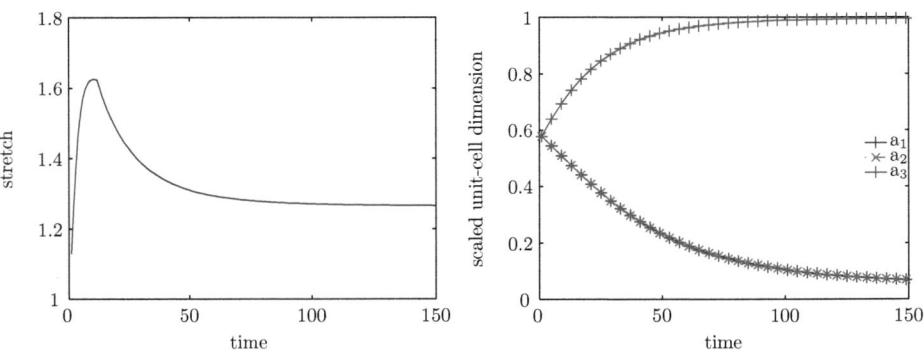

Fig. 5.11: Evolution of the tendon stretch (left) and the unit-cell dimensions (right) during remodeling.

Fig. 5.10 illustrates the displacement of the tendon during the remodeling process. It is extended during the first steps to its maximal length at about the maximal loading. Thereafter the reorientation of the fibers by changing the unit-cell dimensions reflects a significant stiffening of the tendon in the load direction. At the final stage the displacement is reduced to about half the maximal displacement. In Fig. 5.11 the evolution of the stretch is plotted over time on the left hand side. Clearly evident is the stiffening modulated by the exponential evolution equation. On the right side the evolution of the unit-cell dimensions scaled by r_0 are plotted against the time. Starting from an isotropic configuration with $a_i = 1/\sqrt{3}$ the cube length a_3 rises whereas both other dimensions diminish. Finally, a transverse isotropic configuration is reached.

Fig. 5.12 illustrates the fiber configuration for the initial isotropic stage (left) and the final transverse isotropic stage (right). At the initial stage the fibers form a regular cube whereas at the final stage they are all aligned in longitudinal direction. The results of this example resemble perfectly the reported ones by Kuhl and Holzapfel.

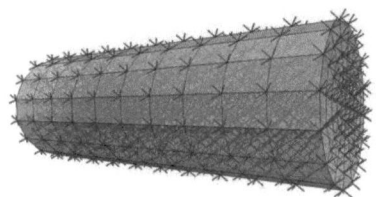 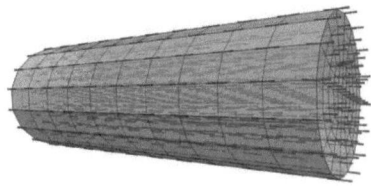

Fig. 5.12: Fiber configuration within the unit cell at the initial configuration (left) and the final configuration (right).

5.5.4 Idealized Artery Model

A final numerical example again considers an idealized tube-like artery subject to uniaxial stretch and internal pressure. This shall serve the demonstration of shortcomings of Kuhl's original remodeling strategy and the effect of our proposed advancements (see Section 5.4). To this end, we follow the example of the tube-like artery presented by Kuhl and Holzapfel [185].

A similar structure is already presented in Section 5.5.1, but the underlying material law and remodeling strategy here is quite different and we do not want to quantitatively compare these two examples. However, the qualitative fiber pattern of an increasing angle with respect to circumferential direction across the thickness should be recovered by this remodeling approach as well.

The applied parameters are summarized in Table 5.3 where we mostly follow the reference parameters in the paper. Unfortunately, the quantitative value of the internal pressure loading is missing in the paper and we have chosen a pressure of 9.6 GPa applied together with the axial stretch incrementally in 25 load steps and hold constant for another 25 steps. Note that we have chosen the same unit system as in the previous example, whereas the paper reports no units. Following the paper we have discretized the structure with eight elements across the thickness and 12 elements in axial direction. In contrast to the displacement-based elements reported in the paper we employed **EAS9**-elements. In circumferential direction we simulated a

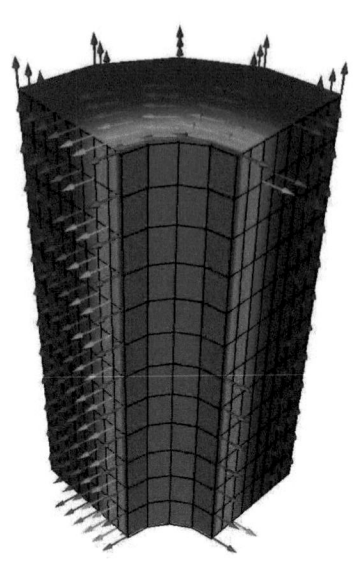 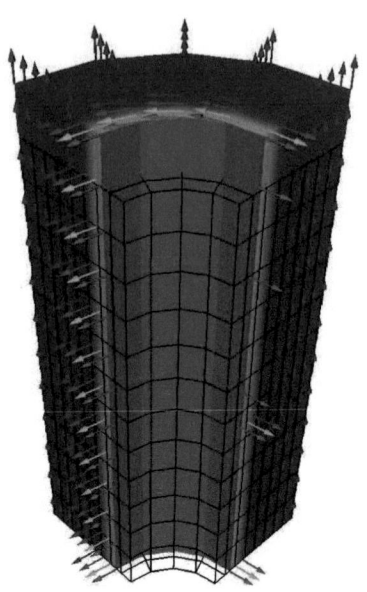

Fig. 5.13: Displaced artery model at first (left) and final (right) step with colors and arrows indicating the first principal stress. It is evident that at the initial step the circumferential stress dominates a large part of the artery, whereas at the final state the circumferential load is carried through the inner part due to the strong remodeled fibers.

quarter of the tube with four elements in circumferential direction and corresponding symmetry boundary conditions.

The general remodeling strategy follows in principle Kuhl's approach described in Section 5.3.3, but with the advanced remodeling evolution discussed in Section 5.4. Thus, the unit-cell dimensions evolve following the ratio of the positive stress eigenvalues (see Equation (5.35)) and the initial chain length r_0 is maintained by rescaling according to Equation (5.34). We always start from an isotropic initial configuration with the cubic network unit-cells arranged circumferentially.

In Fig. 5.13 the initial discretized structure is depicted and the displaced structure is illustrated for the first and the last load step. The colors and arrows represent the maximal principal Cauchy stress and direction. The qualitative difference between the first and last load step is the distribution of circumferential and axial eigenvectors. In the first step with only a 25th of the loading and yet an isotropic material configuration the circumferential load dominates across a major part of the thickness. At the final step with the full loading the pressure load is carried through just about half the structure due to circumferential stiffening resulting from the remodeling.

The remodeling process is illustrated in Fig. 5.14 where on the left the development of the

Table 5.3: Parameters of idealized artery.

	Parameter	Value
Geometry	length	8.0 mm
	inner radius	1.0 mm
	outer radius	3.0 mm
Material	λ	27.293 GPa
	μ	3.103 GPa
	k	1.381×10^{-23} JK^{-1}
	θ	310 K
	N	7×10^{21} m^{-3}
	L	1.594 [-]
	A	1.365 [-]
	r_0	1.0 [-]
	τ_{rem}	0.25 [-]
Loading	p_0	9.6 GPa
	Δl	0.8 [-]
Initial unit-cell	isotropic	$r_0/\sqrt{3}$

cell dimensions for one Gauss-point per element is depicted. Dimension a_3 continuously grows, a_1 continuously decreases and a_2 has distributed values through the thickness. The rescaling to maintain r_0 is observed as a_3 raises higher if both a_1 and a_2 diminish, whereas in the case when both dimensions a_3 and a_2 stay at a reasonable value they are both bound to maintain r_0. In Fig. 5.14 on the right the resulting fiber angle at the final step is plotted versus the thickness. The S-shape already obtained in the example of Section 5.5.1 is evident, however the angle ranges until about 60°. This depends strongly on the ratio of internal pressure and axial stretch and the example is restricted only to qualitative results.

The remodeling is further exemplified in the picture series of Fig. 5.15. Here, just one section of the model consisting of one element row through the thickness is depicted. The material unit-cells are illustrated with colors representing the cell dimension. Starting from unit cubes the remodeling process drives the dimension in thickness direction to zero (blue color). The distribution of the other two cell dimensions gradually develops from uniaxial in circumferential direction at the inside of the tube to orthotropic at the outside.

In the following we want to comment on the results presented by Kuhl and Holzapfel. When applying their original evolution equation, the resulting fiber pattern also represents an alignment with an increasing angle through the thickness with respect to the circumferential direction. However, we were not able to reproduce such results in our implementation. As already stated, the decisive question is the interplay between the stress resulting from the axial stretch and the internal pressure. Admittedly, the quantitative pressure value is not reported

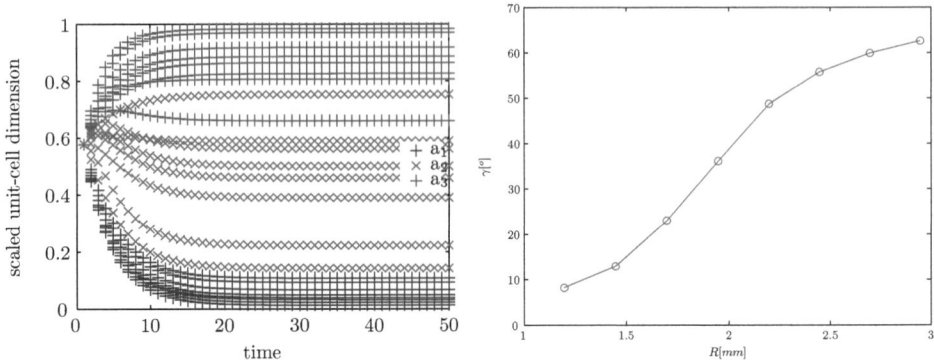

Fig. 5.14: Evolution of the unit-cell dimensions (left) and resulting fiber angle (right).

in the paper, but from locking at the deformed state it seems to be relatively high. However, in our explorations, even a very high internal pressure does not reproduce the desired fiber pattern. In contrast to their example, we initiate remodeling typically with an isotropic fiber pattern. However, changing to a random initial configuration has not improved the results with respect to their reference.

The necessary stress state within the tube which would create the desired fiber pattern could not be recovered with their original remodeling evolution law (see the discussion in Section 5.4 and Fig. 5.3). Here, the problem of the remodeling evolution proposed by Kuhl and Holzapfel becomes obvious. To decrease the cell dimensions, a pressure state is necessary which is only present at the bulge. Everywhere else in the structure the tension stresses in axial direction lead to growing cell dimensions. This is confirmed by the fiber pattern that we obtain with their proposed evolution equation. In Fig. 5.16 the evolution of the cell dimensions is plotted over time. Most of the Gauss-points through the thickness sustain tension in axial and circumferential direction and therefore grow equally. Just at one Gauss-point the axial pressure reduces the axial cell dimension.

This is also observed in the sketched cell dimensions of one element row through the thickness, depicted in Fig. 5.17. Only at one single Gauss-point emerges a magnified circumferential fiber direction (red color). As demonstrated above, this deficit is overcome by our suggested advancement of the remodeling evolution.

5.6 Discussion of the Presented Fiber Remodeling Approaches

Recapitulating the present chapter we would like to discuss the value of considering remodeling approaches in structural modeling of the arterial wall. It is an inherent characteristic of living nature to optimize functionality and thus continuously change and grow. The arterial wall in particular remodels to adopt to its environment, among others in response to structural

5. The Biomechanical Phenomenon of Remodeling 137

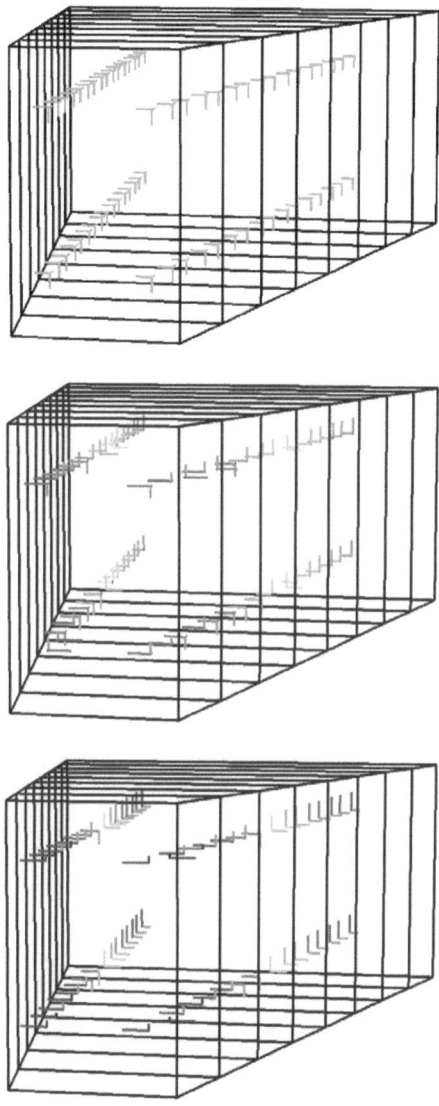

Fig. 5.15: Cutout section consisting of one element row through the thickness of the artery model with sketched material unit cells at each Gauss-point. At the initial step (top) all cells represent cubes, where at the fifth step (middle) and at the final step (bottom) the remodeling of the unit cell dimensions is apparent.

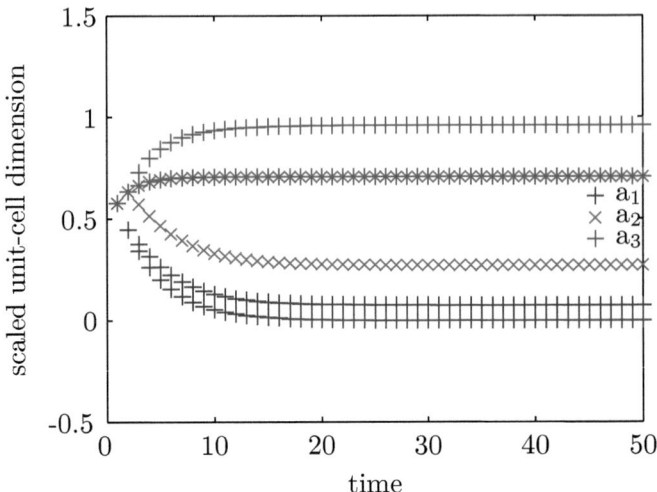

Fig. 5.16: Evolution of the unit-cell dimensions with the original remodeling strategy proposed by Kuhl and Holzapfel.

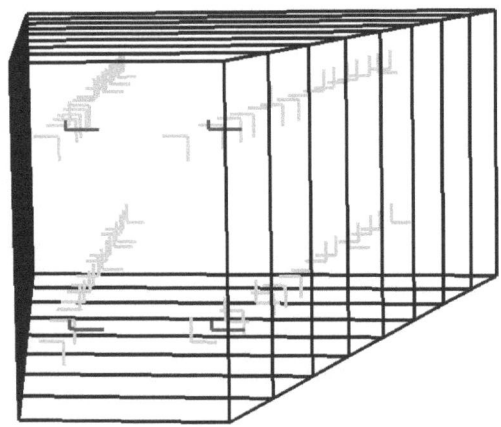

Fig. 5.17: Fiber pattern for one element row due to the original remodeling strategy by Kuhl and Holzapfel.

demands. As such the phenomenon of remodeling is central when modeling the arterial wall.

We have elaborated the underlying equation framework in Section 5.2. It revealed that a purely mechanical remodeling approach is thermodynamically inadmissible and other aspects such as chemo-mechanical or biological phenomena have to be taken into account that drive growth and remodeling. However, the resulting complex and fully coupled system is beyond today's simulation capabilities, the relationship between mechanical forces and biological response is not yet understood in detail, and necessary knowledge about the boundary conditions is still missing.

Therefore, the present approaches to capture remodeling in arterial walls are not aimed at considering the phenomenon in its entire complexity. Rather they shall serve to better understand the global functioning of structural adaptation of the wall. To this end, an idealized tube-like artery is a reasonable reference structure. As demonstrated in the previous sections, the main characteristic of an increasing fiber angle through the wall thickness can usually be captured by the proposed models. We have shown however, that Kuhl's model requires slight modifications. Particularly, setting the principle stresses in relation to each other seems to be necessary to arrive at a smoothly increasing fiber angle.

Nevertheless, there still remain some challenges in trimming the remodeling procedure to obtain such a realistic physiological situation. First of all, the remodeling strategy strongly depends on the underlying anisotropic constitutive model to allow fiber reorientation at all. The corresponding parameters have to be defined a priori which in turn are usually obtained from tissue experiments. This tissue - however - has undergone remodeling already prior to extraction, resulting in a classical dilemma of parameter fitting. In this regard we favor the continuum molecule chain model employed in Kuhl's approach over Hariton's approach since the material parameters have a clear physical meaning.

Second, the underlying structural response needs to be modeled correctly in order to obtain correct remodeling results. As depicted in Fig. 5.6 the resulting increasing alignment angle through the thickness strongly depends on the structure's bulk modulus. Therefore, modeling incompressibility is prerequisite. This draws attention to the involved finite element technology as incompressibility poses high demands on the solution technique. The advanced elements presented in Chapter 3 are well suited to fulfill this demand.

Finally, problems involving more complex geometries and boundary conditions such as the presented carotid bifurcation (see Section 5.5.2) induce bending states. These raise further locking issues and a correct modeling is even more delicate. Clearly, our proposed advanced elements are highly efficient in preventing locking in these circumstances.

Despite these challenges which clearly need to be further pursued in future research, these approaches are well employable to advanced structural models of the arterial wall. Since the tissue's microstructural fiber pattern is not yet obtainable from medical image technology, our purely mechanical approach seems reasonable to enhance the accuracy of the arterial wall structural model. This applies especially to large and complex patient-specific geometries. For instance, prior to simulating the structural response to a specific blood impact, an initial remodeling step is proposed to obtain a reasonable fiber pattern within the wall. This research direction warrants the presented investigation of today's remodeling approaches. We will present applications of remodeling within patient-specific geometries in the following chapter.

6. Simulation of Patient-Specific Arterial Wall Models

One of the key advantages of the finite element method is that highly complex geometries can be analyzed. The aim of biomechanical research of vascular problems is the understanding of diseases. Therefore patient-specific data should be taken into consideration to study inter-individual implications. A major goal of the present thesis is to design a toolchain to efficiently simulate patient-specific problems. This was an important factor for the selection of the methodologies described in the previous chapters which are all well suited for this task. The chapter at hand describes the proposed toolchain in more detail. The issue of generating a reasonable wall geometry is adressed and the proposed approach is further optimized to comply with the wall physiology. Two patient-specific examples considering the aortic arch and the iliac bifurcations are presented to demonstrate the performance of the proposed modeling strategy and the implications of the single methodologies on the simulation are discussed.

6.1 Wall Model Generation

This section describes the generation of our proposed patient-specific arterial wall models. Geometric as well as structural features are included based on segmentation data. To create a feasible and detailed finite-element model, a couple of steps are necessary and several issues have to be resolved or circumvented. The major problem is that the arterial wall itself is barely accessible by today's medical image processing techniques. This makes determination of the arterial wall geometry, especially wall thickness extremely difficult. Therefore, in our proposed arterial wall model we generate the wall by extruding the clearly identifiable blood vessel lumen. The extrusion thickness can either be defined based on histological literature knowledge, or in an enhanced model it can be evaluated on a fixed radius-to-thickness ratio based on additional centerline information. The extrusion of the vessel lumen works on a finite-element mesh of the underlying lumen which allows subsequent analyses by fluid-structure-interaction simulations with a matching interface. The details of the proposed model generation are discussed in the following.

6.1.1 Segmenting and Processing CT-Data

The underlying data basis for our models are always patients' computer tomographic scans. Diagnostic images provide 2D information based on grey-scaled pixel data for each slice. Usually, contrast agents are used to enhance contrast and thus simplify identification of the region of interest. This is quite easy for blood vessels, however other regions such as the soft tissue of the arterial wall or other organs such as the lung are much more difficult to identify. The data is processed by the commercial software package *Mimics* (Materialise, Leuven, Belgium). A three-dimensional representation of the region of interest is obtained by a set of CT image slices.

Several techniques are available to segment the region of interest, such as contrast threshold and region growing algorithms. As mentioned, the lumen of blood vessels are easy to obtain due to the high contrast generated by contrast agents. The resulting region masks are cut at inlet and outlets of the region of interest and usually smoothed to create a three-dimensional representation of the blood vessel lumen body. This pixel-based representation is transferred into a facet-based geometry in the STL format. We refer to the literature or the manual of the software package for details concerning the segmentation and export into facet-based geometries.

The next step consists of the mesh generation of the lumen body, where we apply the commercial meshing software *Harpoon* (Sharc Ltd, Manchester, United Kingdom). The software is capable of handling complex geometries such as the STL representations of blood vessels. It generates hexahedral dominated mixed meshes consisting of predominantly hexahedrals but also of tetrahedrons and five noded pyramids. The benefit of this mixed meshing technique is that the mesh quality of the specific element types themselves is relatively high whereas a pure hexahedral mesh for such complex geometries would be highly distorted, if obtainable at all. Via local refinement zones the overall mesh resolution and quality can be adopted. If subsequent fluid-structure-interaction simulations are intended a sufficiently fine boundary layer mesh can be generated as well. Fig. 6.1 depicts the three steps of segmentation, the corresponding facet-based geometry and a resulting mesh for a human aortic arch.

6.1.2 Generating the Wall Geometry

The arterial wall itself, especially the limits of the outer wall in relation to the surrounding tissue, can usually not be identified via medical image techniques. Even the determination of the in-vivo thickness is difficult. For instance, if only isolated segments are studied the retraction of the segment causes the arterial wall to thicken and the same happens due to removal of the distending pressure (McDonald [210]). For our simulation of patient-specific arterial wall models we therefore generate the wall geometry by extruding the 'wet' surface of the vessel lumen.

The extrusion is performed via an in-house algorithm taking the finite element mesh of the lumen as basis. This has several advantages, as for example the surface mesh quality

6. Simulation of Patient-Specific Arterial Wall Models

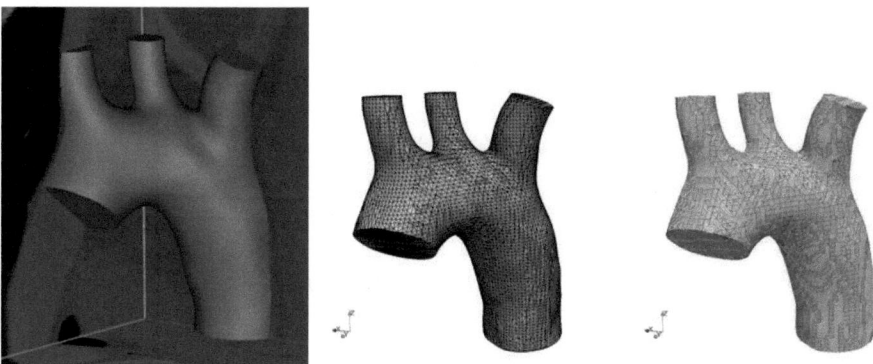

Fig. 6.1: Segmented blood vessel lumen of CT dataset (left), corresponding facet-based geometry (middle) and resulting finite element mesh (right). Here, edge colors indicate hexahedral, tetrahedral, pyramid, and wedge shaped elements.

is already regulated by the meshing tool. Another benefit comes into play when subsequent fluid-structure-interaction simulations are performed, because a matching interface mesh is automatically provided.

The extrusion algorithm takes an averaged nodal director resulting from the mean outside normal direction of the attached surface elements and generates the structural arterial wall elements with a predefined thickness. The wall can be modeled as a number of layered elements with equal thickness, or the extrusion algorithm can be performed several times with different thicknesses whereby each time the nodal averaging results in a smoothing of the corresponding surface. The generated three-dimensional structural wall model consists of hexahedral and wedge-shaped elements. Depending on the resolution of the original lumen surface and the wall thickness the elements become considerably thin. The application of solid-shell elements as described in Chapter 3 is recommended to prevent underestimation of displacements resulting from locking defects. We also refer to the remarks in Section 3.3.1 about element accuracy in mesh layouts resulting from such extrusion methods.

Although this approach seems rather simple and effective there remain several issues, especially if large complex geometries are considered. One is related with the cut regions defined as inlet and outlets of the vessel body. Extrusion of these boundaries by the described averaged director typically results in skew surfaces. However, to apply meaningful boundary conditions such as symmetry these surfaces often need to be flat. Therefore, we map the corresponding nodes back onto the surface defined by the original cut lumen surface. Examples for different results are depicted in Fig. 6.2.

Another major problem arises in bifurcation regions. Depending on the element size h of the surface element, the extrusion thickness t and the bifurcation angle α a twisting of the extruded elements is inevitable, resulting in a irregular Jacobian mapping. The situation

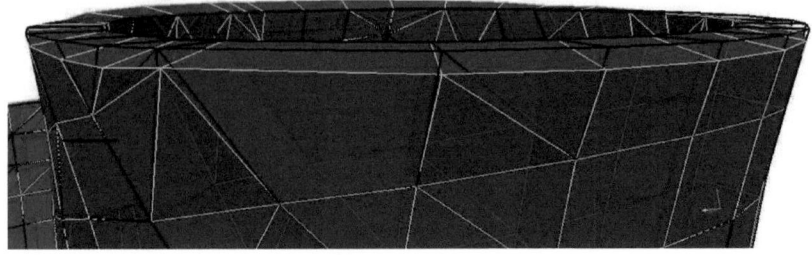

Fig. 6.2: Problematic outlet region resulting from extrusion by an averaged nodal director: original surface (red with white edges) and flattened surface (black edges) mapped onto original cutting plane.

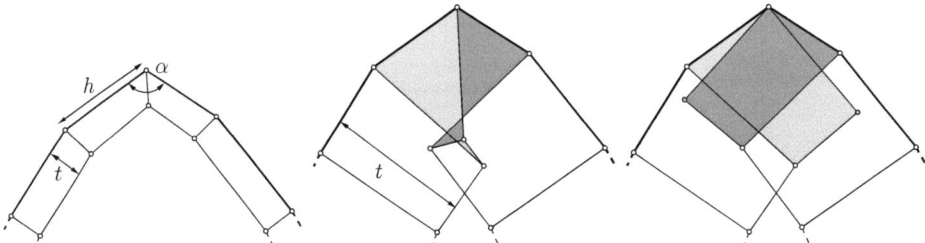

Fig. 6.3: Sketch of regular extrusion case (left) and problematic case (middle). Depending on the parameters t, h, and α the extruded elements might become irregular resulting in an unfeasible mesh. Resolved mesh (right) by introducing a new node and separating the extruded surface.

is sketched in Fig. 6.3. The physiological in-vivo structure shows probably a considerable thickening of such regions, however, as long as the structure and geometry of such regions cannot be determined and processed in a structural finite-element model we decided to tackle this issue as follows to ensure a feasible structural mesh. If such twisted elements are identified during the extrusion algorithm, for example by evaluating the Jacobian mapping at the element nodes, we separate the outer surface nodes, reevaluate the director without taking the problematic neighbor into account and create two physically overlapping finite elements, as sketched in Fig. 6.3. In case of layered extrusion the whole stack is repaired. This results in a consistent finite element mesh. However, a structural weakening of this particular region is evident. But as such problematic regions are quite local we assume that the global structural response of the wall is not drastically disturbed. This certainly needs a closer investigation in future research, especially in quantitative analyses at these regions.

6.1.3 Integrating Centerline Data to Enhance the Wall Model

To achieve a more realistic wall model we enhance the model described so far by introducing further information gained from the centerline of the vessels. The centerline connects the calculated centers of each vessel. We refer to the work of Antiga and colleagues [8, 7] and the references cited therein for further information about methods and algorithms to calculate centerlines. Among others Wolters *et al.* [332] make use of centerlines to construct their FSI-models. We use the centerline tools that are included in the *Mimics* software version 12.1 which allow the calculation of the centerline itself and additional measurements as curvature and fitted diameters. With the centerline data at hand several improvements are obtained, as described in the following.

Boundary conditions First, the inlet and outlets of the vessels can be cut orthogonal to the centerline. This allows a more precise application of a symmetry condition for the arterial wall, but becomes more important when the blood flow is taken into account and corresponding inflow and outflow conditions need to be specified.

Fiber pattern Another enhancement for our proposed wall model based on the vessels centerline is related to the anisotropic fiber-reinforced material model. As discussed in Chapter 4 these material laws are based on the definition of the fiber direction in terms of corresponding unit-vector fields. Within our in-house preprocessing tool we relate these fiber directions to the centerline of the vessel via a local coordinate system at each finite element centre. Therefore the closest centerline point to each finite element is determined and the local axes are related to the centerline neighbor point as axial direction, the vector between element centre and centerline point as radial direction and a third orthogonal direction, see Fig. 6.4.

With this approach complex geometries with several bifurcations and varying vessel orientations can be accurately modeled with a fiber pattern relative to the centerline. Fiber angles are either taken from histological examinations found in the literature or a subsequent remodeling strategy is employed with a feasible initial configuration. Fig. 6.5 is a picture of a fiber pattern for an iliac bifurcation where the fiber directions are illustrated as streamlines showing the winding around the vessel lumen.

Varying wall thickness Moreover, the model is enhanced by additional measurement data offered by the centerline algorithm. This includes the diameter at each centerline point obtained from a best fit circle, an inscribing circle, or a circumscribed circle. We use this information for a better approximation of the wall thickness. As mentioned earlier, medical image technology is not yet capable to determine the wall thickness. The vast majority of simulations are thus performed on models with constant thickness, even though the geometry is patient specific, see among other Breeuwer *et al.* [49]. For large segments along the arterial tree this obviously results in strong under- or overestimations of partial segments. Scotti *et al.* [261] demonstrate strong influence of constant versus varying thickness in FSI simulations of AAAs. Instead, we

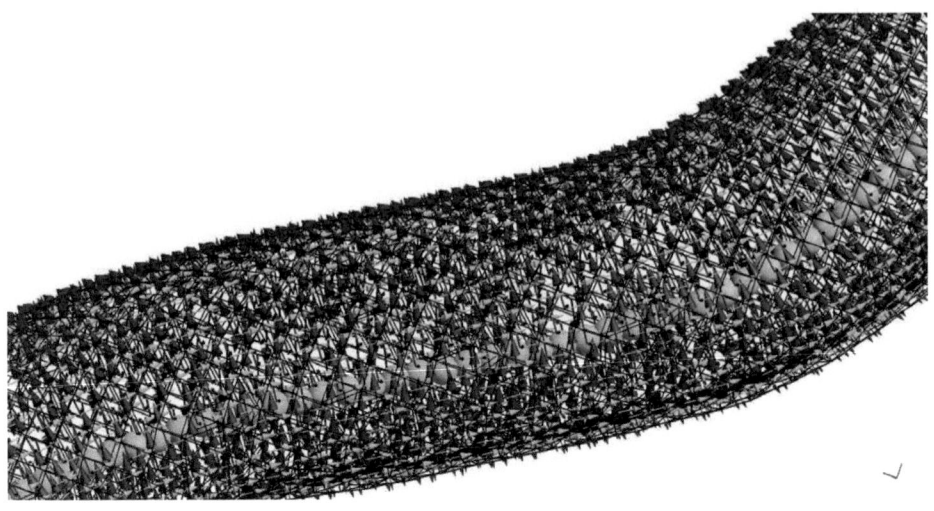

Fig. 6.4: Local element coordinate system related to centerline, red arrows indicate circumferential directions, green arrows axial directions and blue arrows thickness directions.

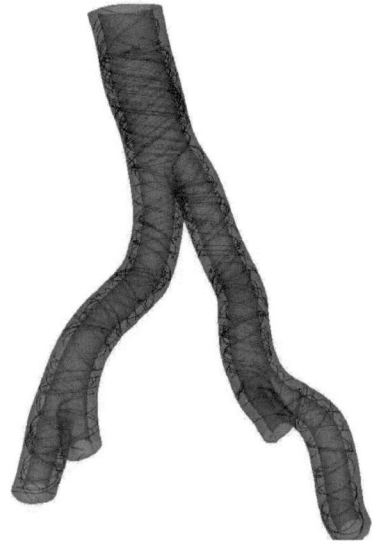

Fig. 6.5: Illustration of fiber pattern at human iliac bifurcation.

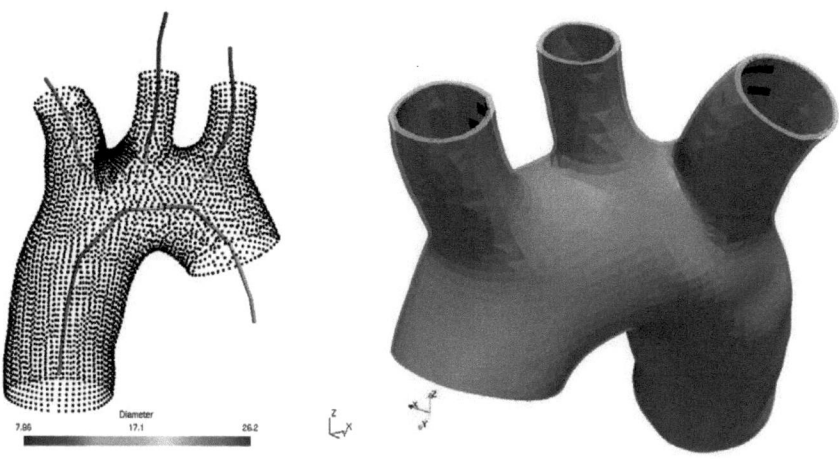

Fig. 6.6: Aortic arch geometry. The color of the centerline points on the left represent computed diameters of the inscribing circles. On the right the resulting extrusion for a constant thickness of 0.8 mm (grey) is compared to an extrusion based on a $R/t = 6.67$ (red).

propose to relate the wall thickness to the vessel radius with a predefined radius-to-thickness-ratio R/t. The local radius is calculated from the centerline data using either the best fit or inscribing diameter. This results in a much more realistic wall thickness along the vascular tree. In Fig. 6.6 on the left the centerline points are depicted for an aortic arch geometry where the color represents the computed diameter. The difference between the aorta and the branching vessels is clearly observable. The right hand side of Fig. 6.6 illustrates the different wall geometries resulting from a constant extrusion thickness of 0.8 mm together with an extrusion based on the varying diameter with $R/t = 6.67$. The overestimated thickness is obvious especially at the outlets of the branching vessels.

Another example considers the extrusion of a segment of the abdominal aorta, the bifurcation into left and right common iliac and the left and right bifurcations into external and internal iliacs. In Fig. 6.7 the difference between an extrusion with constant thickness and based on centerline measurements is demonstrated.

Concluding this section we point out that the described strategies result in advanced and more realistic patient-specific arterial wall models. They fit into the toolchain from image processing to simulation and are computationally efficient. Together with the commercial software packages for segmentation and meshing a fast and widely automatic process of generating patient-specific wall models is provided. Thereby, statistical studies and correlation with clinical data are rendered possible.

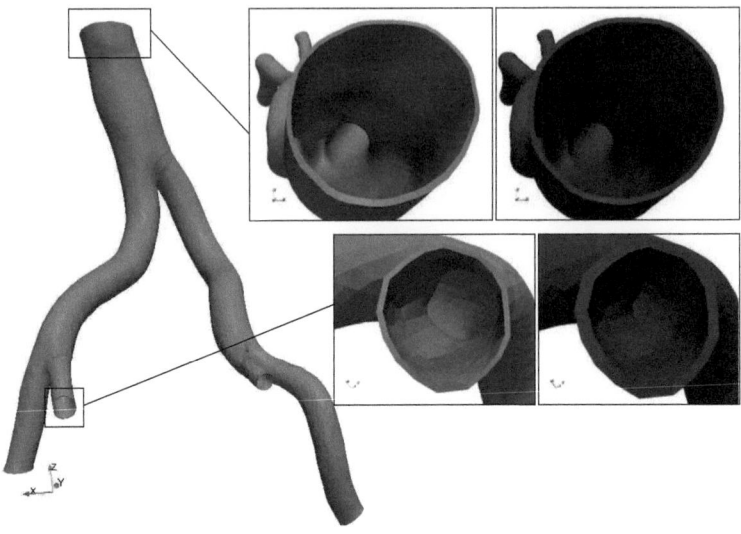

Fig. 6.7: Iliac bifurcation, extruded with constant thickness of 0.55 mm (red) compared to extrusion based on $R/t = 8.33$ (grey).

6.1.4 Considering the Initial Wall Stress State

An important aspect for the assessment of the arterial wall stress is the fact that the unloaded configuration cannot be assumed to be stress-free. Rather, there exists a complex prestress state consisting of axial and circumferential stresses, as discussed in Section 1.2. One aspect of this prestress state is the so-called *spring-open effect*, initially reported independently by Fung [103] and Vaishnav and Vossoughi [313], which describes the effect that arterial rings spring open when sliced in radial direction thereby releasing the prestress.

This prestress phenomenon has to be considered when valuating stress simulations quantitatively and qualitatively. For instance, the *spring-open effect* introduces a bending-type prestress which significantly changes a stress field resulting from solely inflating a tube. In this example of a tube-like artery the superimposed bending and inflation levels the stress field through the thickness. Such a homogenized stress field is assumed to represent an optimized and thus physiological configuration.

In the following, we perform a small numerical example to assess the bending-type prestress representing the physiological state of a healthy artery. Therefore, a stress-free open circular section representing an idealized cut arterial ring is subject to a specified bending load closing the ring as depicted in the illustration of Fig. 6.8. The resulting prestressed structure may subsequently be loaded with physiological blood pressure. Due to singularities and constraints a direct application of closing Dirichlet-conditions usually results in a wrong stress field (Balzani et al. [20, 17]). We therefore apply a different strategy where we introduce a 'virtual gap'-body

6. Simulation of Patient-Specific Arterial Wall Models

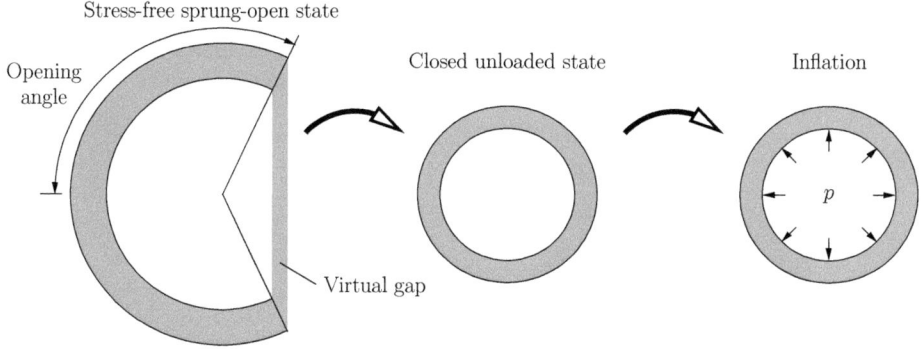

Fig. 6.8: Illustration of *Spring-open effect*.

between the open sections of the ring whose volume is continuously decreased to zero. This additional constraint is solved via a Lagrangean multiplier approach.

The results of a typical arterial ring with an opening-angle of 115°, dimensions and material parameters taken from the idealized carotid artery of Section 5.5.1 are depicted in Fig. 6.9. On the right hand side the initial open ring is also sketched. To demonstrate the importance of advanced element technology to correctly simulate such a bending-type deformation we perform two analyses employing standard displacement-based elements and our proposed Sosh8 solid-shell element. Comparing the results of the first principal Cauchy-stress in the closed configuration, a significant difference is observed in the stress-field between the simulation with standard displacement-based elements on the right and Sosh8-elements on the left. In the displacement-based solution the tensile stress is overestimated by about 200% and due to overestimated pressure stress the overall range is even three times as high. Note that the involved load-bearing fibers contribute only in tension. The locking defects of displacement-based elements are obvious and may lead to a decisive misinterpretation of the stress field within the arterial tissue.

Unfortunately, it is currently impossible to capture the complex prestress in patient-specific simulations. Delfino *et al.* [64] take prestress into account by generating a 'sprung-open carotid bifurcation model' which is first closed and then pressurized. However, this is based on experimental examinations of excised specimen and is therefore not transferable to other patient-specific geometries. Alastrué *et al.* [1] apply a simplified initial strain field obtained from an opening-angle experiment to more complex patient-specific geometries of in vivo CT data. Although this seems to be a valuable approach to include prestress in patient-specific geometries, it is probably limited to tube-like parts due to convergence issues and unavailability of experimental data.

A totally different approach of including prestress is based on growth and remodeling within a temporal multiscale framework, as proposed by Humphrey and Taylor [156] (see also Chapter 5). Including realistic prestress in arterial wall models certainly of substantial importance and

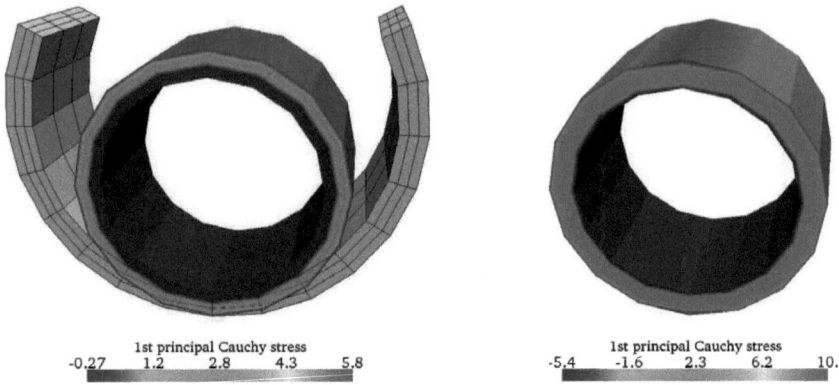

Fig. 6.9: Simulation of the *spring-open effect* in an arterial ring. Comparison of resulting stress field between application of Sosh8-elements (left) and displacement-based elements (right).

thus an interesting topic in future research. Nevertheless, application of any such method requires a correct assessment of the stress response even in complex bending modes. Avoiding spurious locking effects is of major importance and our proposed element set involving advanced element technology represents a highly efficient solution to this issue.

6.2 Simulation of a Patient-Specific Aortic Arch

We consider a human aortic arch with the bifurcating brachiocephalic, left common carotid and left subclavian arteries. The acquired CT data belongs to a 46 year old male showing no obvious disease or lesion at the considered region. Thubrikar [301] presents a finite-element analysis of a realistic aortic arch geometry and Zhang *et al.* [339] present a FSI simulation of this region; however both employ a quite simplified wall model limiting the accuracy of their results.

The vessel lumen of our model are segmented and exported into a facet-based geometry as described in the previous section. A reasonable mesh is generated where we have put a refinement zone at the left subclavian bifurcation to capture the sharp corner. The aortic wall is generated by extruding the lumen with a fixed radius-to-thickness-ratio related to centerline measurements for the inscribing circles. We apply an internal pressure of 4 kPa which corresponds to the difference between the mean systolic and diastolic pressure of 120 mm Hg and 90 mm Hg, respectively. It is argued that the CTA images are taken at the minimal (diastolic) blood pressure. Therefore the loading on this geometry is at most the specified pressure difference to the maximal (systolic) pressure. We emphasize that this does not reflect the realistic physiological stress state in the artery, because the residual stresses within the arteries are

6. Simulation of Patient-Specific Arterial Wall Models

not taken into account. A further limitation might be the missing arterial tethering of the surrounding tissue.

As no experimental data of our employed anisotropic material law was available for the considered region of the aortic arch we use the parameters from Hariton et al. [122] which were however fitted to a human carotid. The aorta is known to be quite stiff; therefore we scaled the parameters by a factor of 1.5. This clearly represents a restriction to the presented simulation. For future quantitative studies experiments and parameter fitting are inevitable.

Nevertheless, with this example we want to study the influence of several parameters of the proposed wall model to the structural response. We therefore assume a specified parameter set as reference and change step by step some of these parameters to investigate their influence. We point out that still a large portion of the model parameters are preliminary assumptions with respect to the true physiological situation and therefore we do not claim that the results are quantitatively correct. We merely aim to single out the qualitative influence of several model parameters.

6.2.1 Problem Setup and Reference Parameters

First of all we present the problem setup, geometric and boundary conditions and material parameters. This should serve as a reference problem from where we vary several conditions in the subsequent influence study. We apply the Holzapfel-material law (Section 4.3.1) with parameters summarized together with further geometric and boundary conditions in Table 6.1. In Fig. 6.10 the mesh and the fiber pattern are illustrated.

Table 6.1: Reference parameters of human aortic arch model.

	Parameter	Value
Geometry	Radius-to-Thickness-Ratio	8.33 [-]
Material	μ	53.61 kPa
	k_1	41.7 kPa
	k_2	20.85 [-]
	κ	600 kPa
	Fiber angle w.r.t. centerline	20°
Bound. Cond.	Internal Pressure	$p_0 = 4$ kPa
	Loadcurve	$1 - \cos(t\pi/2)$, $0 < t \leq 1.0$
	Loadstep	$\Delta t = 0.02$
	Dirichlet Conditions	in- and outflow fixed
Mesh	5256 8-node hex elements	Sosh8-solid-shell see Sec. 3.5.2
	6912 6-node wedge elements	Sosh6-solid-shell see Sec. 3.5.3
	Number of element layers	2

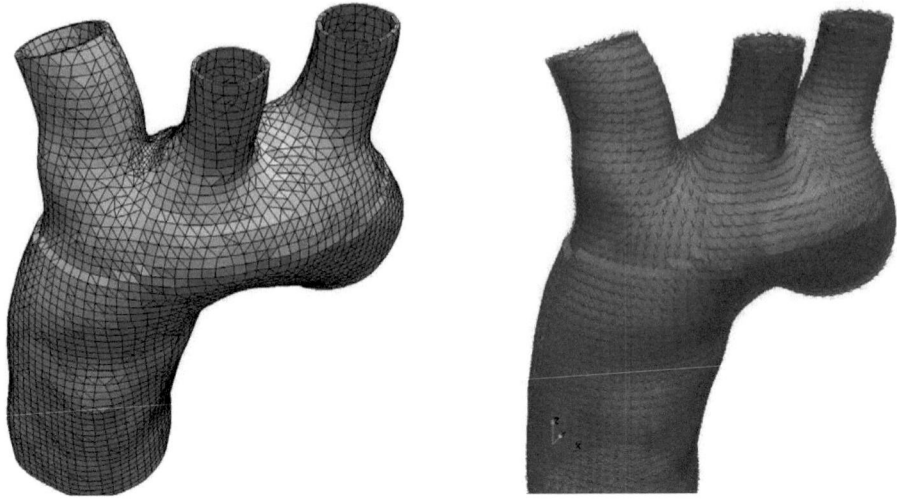

Fig. 6.10: Aortic arch reference example: mesh (left) and fiber pattern (right).

Several simulation results are depicted in Fig. 6.11 and Fig. 6.12. We remark that the depicted stress scale is cut at 200 kPa to exclude stress singularities existent at the ridge of the left subclavian bifurcation where the mesh is repaired at 10 elements according the strategy described in Section 6.1.2.

6.2.2 Influence Study

Element Technology Instead of using the advanced solid-shell elements described in detail in Chapter 3 another simulation has been performed with standard displacement-based elements. The overall difference in displacement magnitude is below 2 % and no locally differing load transfer due to overestimated bending stiffness is observable. One reason for this small difference and thus a minor influence of locking is probably the dominating membrane stress state. It is evident that in the displacement plots of Fig. 6.11 the wall structure is predominantly blown up and no strong bending deformation is present. Also volumetric locking does not seem to play a significant role. Nevertheless, we would like to emphasize that there might well exist a bending-type prestress within the arterial wall, which if taken into account necessitates the application of locking eliminating techniques.

Material Law To estimate the influence of the material we simply raised all parameters by 33%. This yields a reduction in displacement magnitude of 17 %, but the overall deformation pattern is the same and also the stress magnitude is close to the reference simulation. To study the influence of the anisotropy we applied a simple isotropic Neo-Hookean material law with $\mu = 100$ kPa which is roughly the sum of the ground-substance and fiber stiffness of the

6. Simulation of Patient-Specific Arterial Wall Models

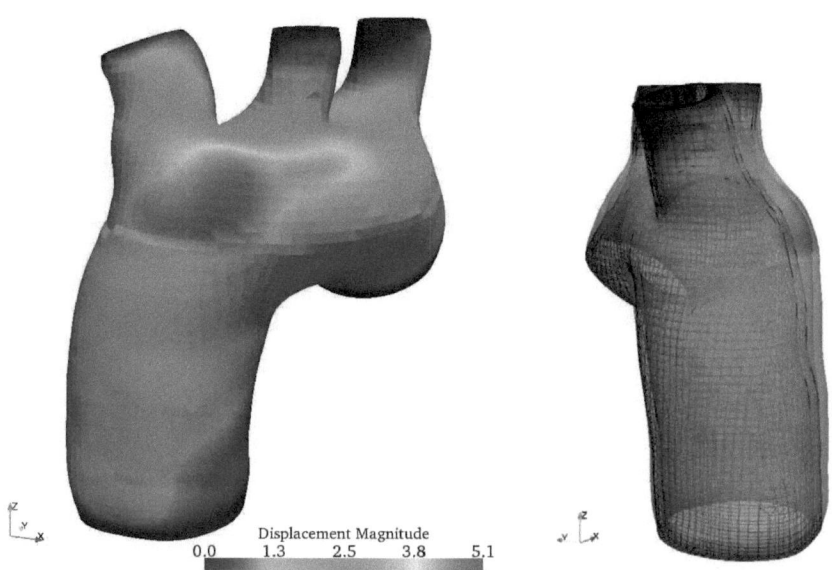

Fig. 6.11: Aortic arch reference results: displaced structure with undisplaced mesh (right).

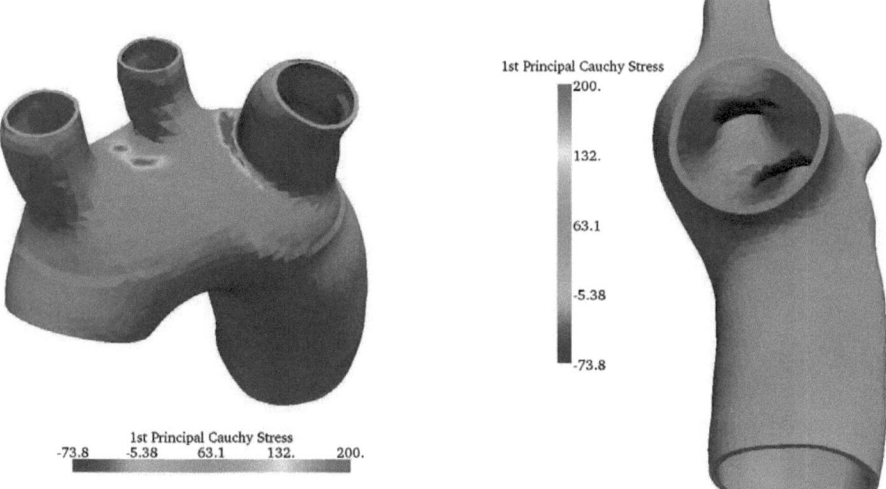

Fig. 6.12: Aortic arch reference results: First principal Cauchy stress.

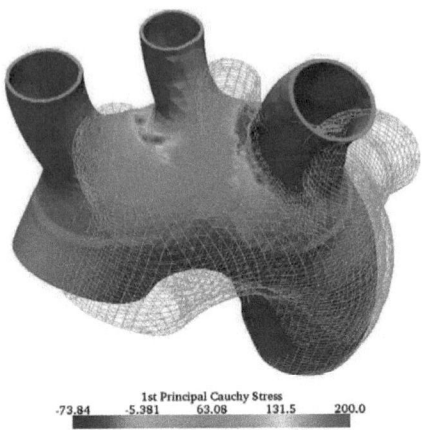

Fig. 6.13: Aortic arch result of boundary condition study: first principal Cauchy stress of simulation with symmetry conditions at all outlets compared to the displaced structure of the reference problem (grey).

anisotropic reference simulation. This results in a 14% magnification of displacements and a slightly smoother but quantitatively equal stress distribution.

Boundary Conditions Changing the boundary conditions from fixed displacements at all in- and outflow boundaries to symmetry conditions at all outflow boundaries changes the global deformation, as depicted in Fig. 6.13. However, the stress state is not significantly different.

Wall Thickness Varying the wall thickness by changing the radius-to-thickness-ratio to $R/t = 6.67$ and thus increasing the thickness by 25% results in a maximal displacement magnitude of 85%. Furthermore, when we switched of the centerline-related extrusion and employed a constant wall thickness of 1.2 mm, the maximal displacement magnitude decreased by 10%. However more importantly, the overall displacement shape changed towards a stronger pressurizing of the aortic vessel, but a smaller displacement of the significantly thickened bifurcating vessels (compare Fig. 6.14).

Fiber Pattern and Remodeling To study the influence of the fiber pattern within the anisotropic material law we first of all increased the number of elements across the thickness to three. We performed a simulation with fixed predefined fiber angles of 15° for the inner two elements (Media) and 45° for the outer element (Adventitia). Furthermore, we applied the remodeling strategy of Hariton (see Section 5.3.2) starting with an initial fiber angle of 20° across the wall. Where the first variation with predefined fiber angles barely changes the result, the remodeling strategy reduced the maximum displacement by 10%. As is visible in Fig. 6.15

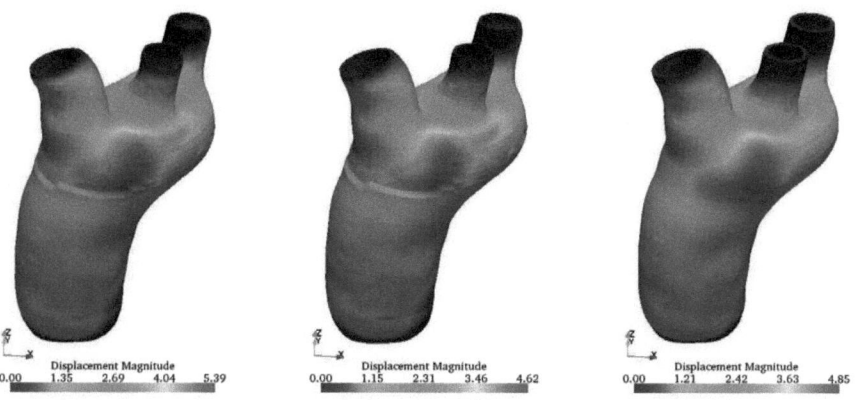

Fig. 6.14: Aortic arch result of different wall thicknesses. Reference result with varying thickness and $R/t = 8.33$ (left), result with varying thickness and $R/t = 6.67$ (middle), and result with constant thickness of 1.2 mm (right).

on the right, the resulting fiber pattern is oriented much more circumferentially, compared to the predefined fiber pattern with 15° and 45° for Media and Adventitia.

6.3 Simulation of a Patient-Specific Section at the Iliac Bifurcations

As a second example we study the aortic section between the abdominal aorta and the external and internal iliacs bifurcations. Again, we are not interested in a quantitative analysis at this stage. Rather, we aim at demonstrating the capabilities of the presented methodologies in such large-scale examples. We stick to the reference parameters of the previous example, see Table 6.1, with the only modification of the fiber angles to 22° for the inner elements and 39° for the outer ones. In addition, a second simulation is run including remodeling. In Fig. 6.16 is the resulting displaced structure depicted. During the rising loadfactor the structure deforms initially orthogonally to the common iliac bifurcation, but then a significant rigid body displacement of the left iliac artery can be observed. This raises the question of the perivascular tethering which would clearly influence such a result.

Regarding remodeling in such a complex patient-specific geometry we subsequently focus on the fiber pattern around the common iliac bifurcation which is certainly a region of a complex mechanical load state. In the first simulation we applied a fixed fiber angle with respect to the centerline of the vessels. In this way we are able to follow the evolving vessels and the fibers wind themselves around the vessels according to the physiologically expected pattern. However, the centerline calculation has difficulties at the bifurcation which results in relatively

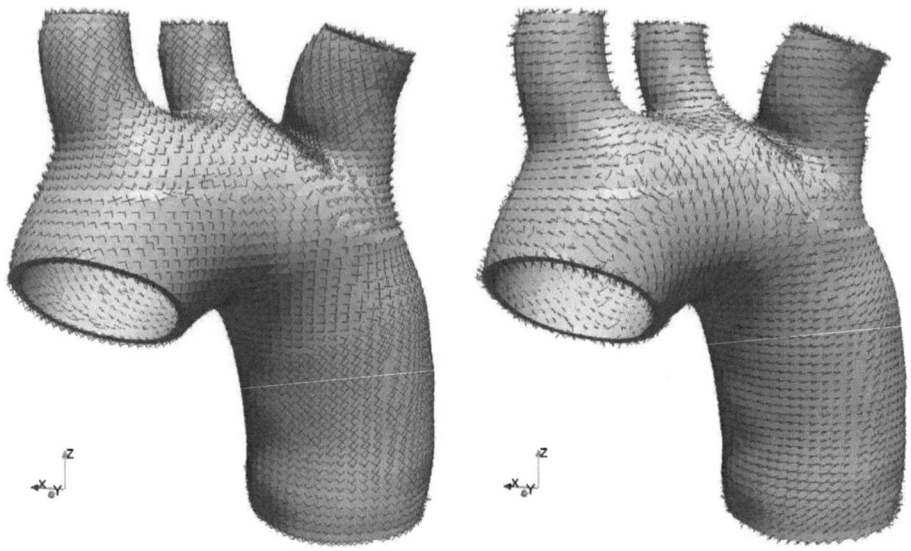

Fig. 6.15: Aortic arch result of different fiber patterns. Predefined fiber pattern (left) with 15° for the inner two elements (Media) and 45° for the outer element (Adventitia). The fiber pattern resulting from the remodeling strategy (right) is significantly different.

large distances between the centerline points. Thus, our proposed algorithm to define local coordinate systems at the elements suffers from this lack of accuracy yielding to a non-smooth fiber pattern at this region, see Fig. 6.17 above. In comparison, the resulting fiber pattern from the remodeling strategy, see Fig. 6.17 below, represents a mostly smooth pattern representing a stiffening fiber bundle running across the apex of the bifurcation. This yields also a smoother stress distribution around that region.

6.4 Discussion

Summarizing this chapter of patient-specific simulations we discuss our proposed arterial wall modeling with its benefits and limitations. The toolchain from patient-specific data to simulation results involves segmentation, meshing, wall generation and simulation. For segmentation and meshing of the vessel lumen we rely on commercial software packages which provide efficient and widely robust solutions. Our developed wall model generation fits well into this semi-automatic process. The automatic and computationally efficient extrusion process takes the provided centerline information into account, ensures the required flat boundary surfaces at inlet and outlets and manipulates complicated sharp angle regions to guarantee a feasible structural wall mesh.

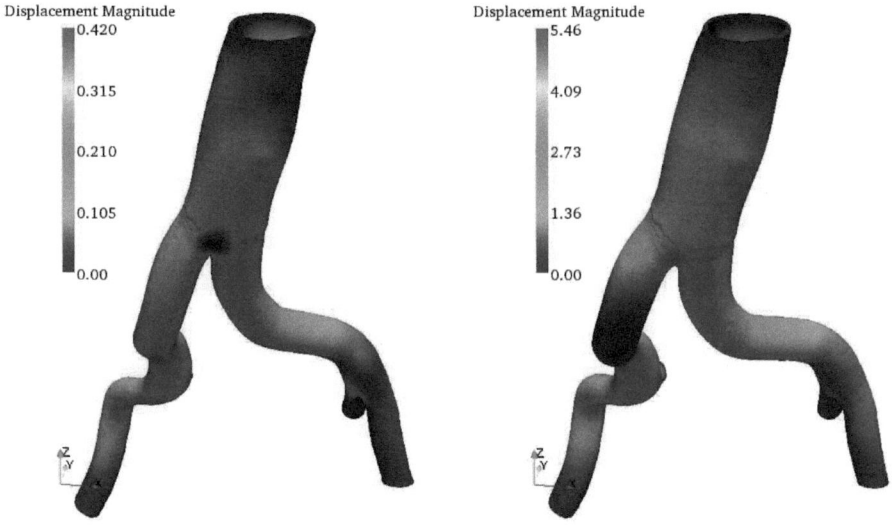

Fig. 6.16: Aortic section at the iliac bifurcations. Displaced structure at 25% loading (left) and full loading (right).

In the present context we have focused on the structural response of such geometries. Two examples employing our proposed toolchain for modeling and simulating patient-specific arterial wall geometries were presented, one reflecting the aortic arch and the other the iliac bifurcations. Simulations of such large complex geometries are still rare. To the authors' knowledge, simulation of such geometries in conjunction with the presented highly sophisticated wall model including varying thickness, anisotropy and advanced element technology has not been considered in the literature yet. Moreover, including remodeling approaches within simulations of patient-specific geometries represents new results within this research field. An evident future analysis step would be to add the impact of the blood flow in a fully coupled FSI-simulation. The presented toolchain is directly applicable for such simulations within the existing in-house finite element code.

Regarding the geometric wall generation the strength of our proposed model lies in enabling a physiologically varying wall thickness along the vascular tree based on a centerline related extrusion algorithm. Our proposed element repair strategy also ensures feasible meshes, especially at bifurcation regions. One limitation of this approach deserves attention though. While the physiological wall at bifurcation regions is characterized by a considerable thickening (Thubrikar [301]), our proposed extrusion algorithm often yields a structural weakening at such regions due to the mesh repairing strategy. This geometry induced dilemma could be resolved generally if a special meshing strategy was employed at such regions. For instance, an automatic

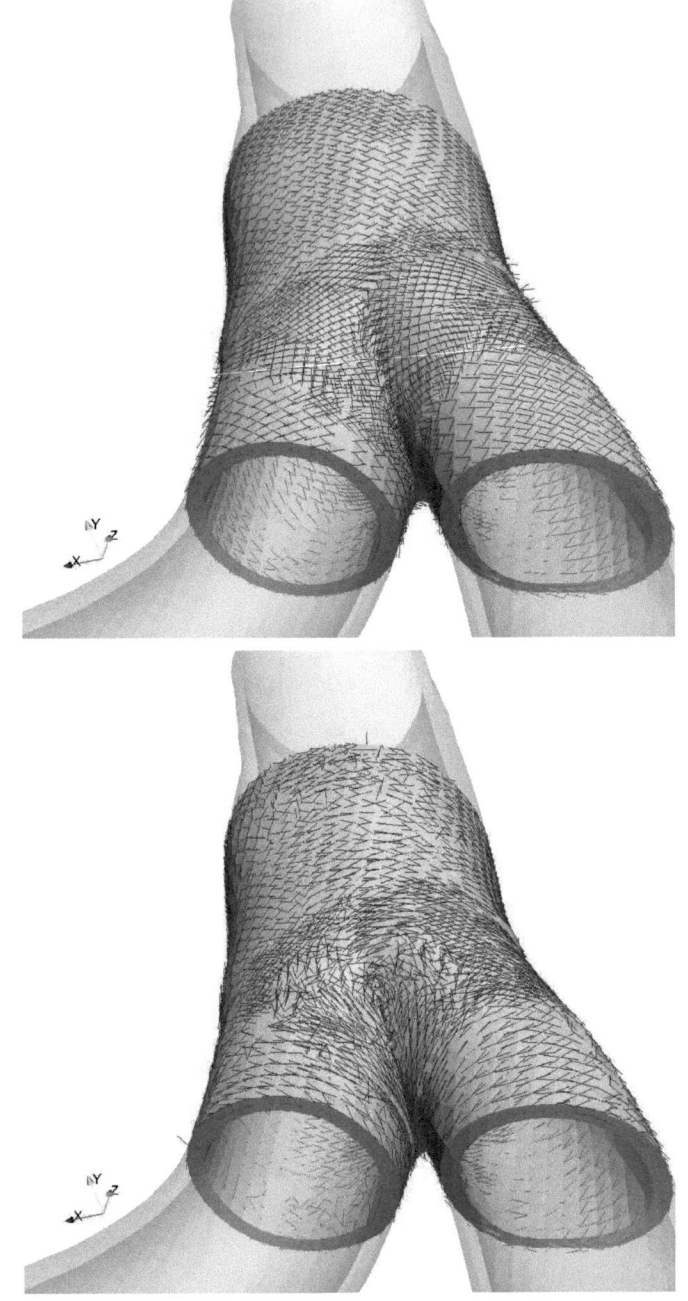

Fig. 6.17: Closeup of fiber pattern around common iliac bifurcation: Constant pattern of 22°/39° (above) and remodeling result (below).

tetrahedral mesh generation could be implemented based on the extruded nodes. However, this would introduce another issue, namely the typically poor performance of tetrahedrons.

Given that arteries are inherently thin-walled structures an adequate simulation of the structural response calls for advanced element technology. Indeed, our analysis of the physiologically inherent bending-type prestress of a healthy artery in the spring-open example revealed the tremendous impact of locking to the resulting stress field. Only with advanced element technology to eliminate such locking phenomena an accurate physiological response could be simulated. Our set of proposed elements proved to solve this issue efficiently.

In contrast, the global displacement results were not significantly influenced by the applied element technology within the presented patient-specific examples. One reason for this seems to be that the major structural response is membrane dominated, which is why the elimination of locking due to bending plays a comparably minor role. It should be remarked though that even in a globally membrane dominated displacement shape the load transfer might be influenced by local bending. Consequently, a detailed analysis of such particular regions will certainly require approaches eliminating locking phenomena. The same applies as soon as the physiological bending-type prestress of a healthy artery would be included into such simulations. In these cases a significant influence of appropriate element technology is to be expected. More generally, there is still room for improving the accuracy of the presented simulations. This includes mesh convergence, dependency of element shape quality and the influence of membrane and volumetric locking which still exists in wedge-shaped elements.

Furthermore, experimental investigation of arterial wall tissue providing parameters for the employed model is certainly of major importance regarding the results of these simulations. In the present context no experimental data for the considered regions was available which is why we relied on rescaled parameters reported in literature instead of fitting the material model to experimental data. This is an important limitation of the presented simulations and a closer relationship to experimentally approved material behavior is recommended for future research. Whether it will be possible to determine the complex anisotropic structure with experimental methods is questionable, especially in vivo. Additional methods in the field of medical image technology could be taken into account, for instance with respect to fiber contents and direction. Noteworthy, the impact of anisotropy was significant in our simulations.

Regarding remodeling we have successfully included one of the remodeling approaches discussed in the previous chapter in the presented patient-specific simulations. The resulting fiber alignment was physiologically plausible, especially in terms of a natural thus optimal load transfer. In addition, remodeling proved to clearly influence the results of our simulations. Employing our purely mechanical approach to determine the unknown fiber pattern in the patient-specific examples yielded a significantly different displacement level compared to the anisotropic law with fixed fiber directions. Further simulation examples should be conducted to substantiate these findings and correlations with experimental results and medical image technology would be desirable.

Finally, we want to address the stability and robustness of the computations and the considered methods. A considerable number of simulations was performed with patient-specific geometries of varying complexity, several different parameter sets, and different computational methods. Thereby, we experienced a few issues of convergence and computational breakups. Each of the involved methodology contributes to these stability issues: the involved element technology is known to have stability limits for large strains (see Section 3.3.3), strong anisotropy comes along with convergence problems (personal communication with Daniel Balzani, 2008), a factor which becomes even worse within a remodeling simulation. Moreover, the issue of incompressibility is computationally difficult and the applied penalty approach is known to induce problems. Finally the complex patient-specific geometries in general feature a more complex structural behavior compared to idealized cylindrical models.

None of these instability factors could be singled out as major cause of instability and computational breakups. Rather the cumulation of issues seems responsible for the observed convergence issues. With the help of computationally beneficial methods, for instance the inclusion of viscosity or other sources of damping, varying boundary conditions, or refined element meshes, these problems could be eased, despite not eliminated. Further research should therefore be directed to improving robustness of such simulations.

7. Conclusion and Outlook

The present contribution was concerned with modeling and simulation of patient-specific arterial wall structures by means of the finite element method. Our focus on precise modeling of the arterial wall in particular was motivated by the hypothesis that the mechanical stress within the wall plays a major role in the onset and development of prevalent vascular diseases like atherosclerosis and aneurysms. The interest in this biomechanical research area has grown tremendously in recent years and quite some progress in understanding hemodynamics, arterial wall characteristics and microstructure, and the correlation between mechanical stress and disease has already been reached. Recent developments of medical image processing and computational analysis capabilities further promoted the analysis of complex biomechanical problems within physiologically realistic, patient-specific geometrical models.

However, such analyses are inevitably three-dimensional, the involved patient-specific geometries are complex and thus require enormous computational efforts, which is why established analysis methods as built into commercially available software packages have so far been limited in their significance for patient-specific application.

The goal of the presented work was therefore to identify, appraise, and adopt state-of-the-art methods within the framework of nonlinear structural finite element analysis, implement them into the sophisticated in-house research code, and finally provide an efficient toolchain for the analysis of patient-specific vascular problems. The comprised methods cover the topics of finite element technology for three-dimensional solids, constitutive modeling suitable for arterial wall tissue, and the computational modeling of the biomechanical phenomenon of fiber remodeling.

Concerning finite element technology, locking defects in three-dimensional solid elements are a key issue in modeling thin-walled biomechanical structures. An important aspect of the present work was thus to employ efficient and reliable methods to eliminate these locking problems with a special focus on so-called solid-shell elements. A set of advanced elements was proposed including a bulky hexahedron, a hexahedral solid-shell, and a wedge-shaped solid-shell. Their performance with respect to locking elimination was evaluated in several benchmark examples, where the newly developed wedge-shaped solid-shell yielded particularly remarkable out-of-plane results.

The microstructure of arterial wall tissue consisting of ground substance, elastin and collagen fibers, as well as smooth muscles exhibits a complex nonlinear material behavior in the large strain regime. To capture the major factors in constitutive equations, demanding models have been presented in recent years. Within this context, the present work provided an overview of the most popular material models, including effects of incompressibility, anisotropy, microstructure, and viscoelasticity. In addition, the corresponding computational algorithms were implemented into the in-house research code. Their effectiveness was demonstrated by a set of numerical examples addressing the essential features of the models.

Finally, the biomechanical phenomena of growth and remodeling were addressed. Based on an in-depth review of the literature on mechanically motivated approaches and the underlying equations, strengths and weaknesses of three popular methodologies are discussed. Focusing on the aspect of remodeling as a change in the microstructure via reorientation of the fiber pattern, we selected two approaches which seemed particularly suitable for large-scale patient-specific problems. Elaborating on these approaches, we proposed some modifications to improve their performance. Applying these remodeling strategies in examples of idealized structures like cylindrical arteries, bifurcations, or tendon models proved successful, since results were fully in line with available literature.

With these three methodical approaches at hand we designed a toolchain to model patient-specific arterial walls to a high degree of accuracy. This involved the problem of generating the geometric wall, employing appropriate finite element meshes, assigning sophisticated anisotropic material models, and evaluating the influence of remodeling strategies. A further advancement over existing patient-specific modeling efforts was achieved by our suggestion to additionally incorporate centerline data available from commercial imaging software; this resulted in a reasonably correct winding of the reinforcing fibers along the vasculature. In addition, we proposed to include evaluated diameter data to obtain a physiologically realistic geometry featuring a varying wall thickness along the vasculature. A valuable benefit for future investigations lies in the fact that the proposed wall model can be directly employed in fully coupled fluid-structure-interaction simulations available with the in-house research code. Future developments may also include coupling to mass transfer.

We thus provide a semi-automatic reasonably efficient computational toolchain to assess arterial wall stress within patient-specific geometries. Nevertheless, there are still a number of limitations involved within this computational model of such a highly complex structure like the vasculature. One of these limitations are missing experimental validations for the proposed models. There are a few examples in the literature where constitutive models are fitted to experimental data. However, the significance of these results still seems limited since the structural behavior is thought to vary substantially between different species, subjects and regions within the vasculature. Experimental setups and protocols are still labor intensive, especially if complex characteristics such as anisotropy are requested. Certainly, capturing material behavior in vivo is desirable, yet this seems to be a rather long-term future goal since already a considerable amount of experimental data is necessary to validate existing models.

Another shortcoming of our model was that the prestress state naturally present in the wall was not considered. Today, there seems to be no methodology available to include the complex prestress state within patient-specific geometries. This generally hinders a quantitative interpretation of simulation results and future research should thus be directed to including prestress

7. Conclusion and Outlook

in such analyses. For instance further experimental investigations may be incorporated or approaches like the temporal multiscale simulation of mechanically induced growth (as envisaged by Humphrey and Taylor [156] as 'Fluid-Solid-Growth' model) could be implemented. So far, these research efforts still seem to be very much in their infancy.

In addition, from a computational mechanics point of view, the proposed methods are affected by a few further shortcomings. Cumulation of demanding features entailed stability and convergence problems which warrant further research in solution techniques. Large strain instabilities in the proposed element technology together with effects of strong anisotropy and incompressibility enforced via penalty methods resulted in badly scaled matrices for the solver. The aspired coupling to the fluid field and mass transfer will likely yield further computational challenges.

The proposed arterial wall model provides a powerful tool to analyze arterial wall stress in patient-specific geometries since it incorporates a comprehensive set of advanced methods. Its power and efficiency has been demonstrated in application to large-scale examples of a human aortic arch and a section around the iliac bifurcation. These patient-specific examples go far beyond data presented in earlier research. In addition, the proposed arterial wall model is also highly suitable for future studies that pool the data of several cases to gain insight into the mechanical environment of the vasculature. An ultimate goal would then be to correlate results with pathological data, to be able to better predict the onset and development of vascular diseases.

Bibliography

[1] V. Alastrué, E. Peña, M. A. Martínez, and M. Doblaré. Assessing the use of the "opening angle method" to enforce residual stresses in patient-specific arteries. *Annals of Biomedical Engineering*, 35(10):1821–1837, 2007.

[2] P. Alford, J. D. Humphrey, and L. Taber. Growth and remodeling in a thick-walled artery model: effects of spatial variations in wall constituents. *Biomechanics and Modeling in Mechanobiology*, 7(4):245–262, 2008.

[3] D. Ambrosi, A. Guillou, and E. Di Martino. Stress-modulated remodeling of a non-homogeneous body. *Biomechanics and Modeling in Mechanobiology*, 7(1):63–76, 2008.

[4] D. Ambrosi and F. Mollica. On the mechanics of a growing tumor. *International Journal Of Engineering Science*, 40(12):1297–1316, 2002.

[5] U. Andelfinger. *Untersuchungen zur Zuverlässigkeit hybrid-gemischter Finiter Elemente für Flächentragwerke*. PhD thesis, Institut für Baustatik, Universität Stuttgart, 1991.

[6] U. Andelfinger and E. Ramm. EAS-elements for two-dimensional, three-dimensional, plate and shell structures and their equivalence to HR-elements. *International Journal for Numerical Methods in Engineering*, 36(8):1311–1337, 1993.

[7] L. Antiga. *Patient-Specific Modeling of Geometry and Blood Flow in Large Arteries*. PhD thesis, Dipartimento di Bioingegneria, Politecnico di Milano, 2002.

[8] L. Antiga, M. Piccinelli, L. Botti, B. Ene-Iordache, A. Remuzzi, and D. Steinman. An image-based modeling framework for patient-specific computational hemodynamics. *Medical & Biological Engineering & Computing*, 46(11):1097–1112, 2008.

[9] F. Armero. On the locking and stability of finite elements in finite deformation plane strain problems. *Computers & Structures*, 75(3):261–290, 2000.

[10] E. M. Arruda and M. C. Boyce. A three-dimensional constitutive model for the large stretch behavior of rubber elastic materials. *Journal of the Mechanics and Physics of Solids*, 41(2):389–412, 1993.

[11] E. M. Arruda, S. C. Calve, K. Garikipati, K. Grosh, and H. Narayanan. *Mechanics of Biological Tissue*, chapter Characterization and Modeling of Growth and Remodeling in Tendon and Soft Tissue Constructs, pages 63–75. Springer, Berlin, 2006.

[12] J. H. Ashton, J. P. Vande Geest, B. R. Simon, and D. G. Haskett. Compressive mechanical properties of the intraluminal thrombus in abdominal aortic aneurysms and fibrin-based thrombus mimics. *Journal of Biomechanics*, 42(3):197–201, 2009.

[13] F. Auricchio, L. Beirão da Veiga, C. Lovadina, and A. Reali. A stability study of some mixed finite elements for large deformation elasticity problems. *Computer Methods in Applied Mechanics and Engineering*, 194(9-11):1075–1092, 2005.

[14] F. Auricchio, L. Beirão da Veiga, C. Lovadina, and A. Reali. The importance of the exact satisfaction of the incompressibility constraint in nonlinear elasticity: mixed FEMs versus NURBS-based approximations. *Computer Methods in Applied Mechanics and Engineering*, In Press, Corrected Proof:–, 2008.

[15] Y. Başar and D. Weichert. *Nonlinear Continuum Mechanics of Solids–Fundamental mathematical and physical concepts*. Springer, Berlin, 2000.

[16] J. M. Ball. Convexity conditions and existence theorems in non-linear elasticity. *Archive of Rational Mechanics and Analysis*, 63:337–403, 1977.

[17] D. Balzani. *Polyconvex Anisotropic Energies and Modeling of Damage Applied to Arterial Walls*. PhD thesis, Institut für Mechanik, Universität Duisburg-Essen, 2006.

[18] D. Balzani, F. Gruttmann, and J. Schröder. Analysis of thin shells using anisotropic polyconvex energy densities. *Computer Methods in Applied Mechanics and Engineering*, 197(9-12):1015–1032, 2008.

[19] D. Balzani, P. Neff, J. Schröder, and G. A. Holzapfel. A polyconvex framework for soft biological tissues. adjustment to experimental data. *International Journal of Solids and Structures*, 43(20):6052–6070, 2006.

[20] D. Balzani, J. Schröder, and D. Gross. Numerical simulation of residual stresses in arterial walls. *Computational Materials Science*, 39(1):117–123, 2007.

[21] K.-J. Bathe, F. Brezzi, and S. W. Cho. The MITC7 and MITC9 plate bending elements. *Computers & Structures*, 32(3-4):797–814, 1989.

[22] K.-J. Bathe and E. N. Dvorkin. A four-node plate bending element based on mindlin/reissner plate theory and a mixed interpolation. *International Journal for Numerical Methods in Engineering*, 21(2):367–383, 1985.

[23] K.-J. Bathe and E. N. Dvorkin. A formulation of general shell elements – the use of mixed interpolation of tensorial components. *International Journal for Numerical Methods in Engineering*, 22(3):697–722, 1986.

[24] Y. Bazilevs, V. M. Calo, Y. Zhang, and T. J. R. Hughes. Isogeometric fluid-structure interaction analysis with applications to arterial blood flow. *Computational Mechanics*, V38(4):310–322, 2006.

[25] N. Büchter and E. Ramm. Shell theory versus degeneration - a comparison in large rotation finite element analysis. *International Journal for Numerical Methods in Engineering*, 34(1):39–59, 1992.

[26] N. Büchter, E. Ramm, and D. Roehl. Three-dimensional extension of non-linear shell formulation based on the enhanced assumed strain concept. *International Journal for Numerical Methods in Engineering*, 37(15):2551–2568, 1994.

[27] T. Belytschko and L. P. Bindeman. Assumed strain stabilization of the eight node hexahedral element. *Computer Methods in Applied Mechanics and Engineering*, 105(2):225–260, 1993.

[28] T. Belytschko, W. K. Liu, and B. Moran. *Nonlinear Finite Elements for Continua and Structures*. John Wiley & Sons, 2000.

[29] P. G. Bergan and C. A. Felippa. A triangular membrane element with rotational degrees of freedom. *Computer Methods in Applied Mechanics and Engineering*, 50(1):25–69, 1985.

[30] P. Betsch, F. Gruttmann, and E. Stein. A 4-node finite shell element for the implementation of general hyperelastic 3d-elasticity at finite strains. *Computer Methods in Applied Mechanics and Engineering*, 130(1-2):57–79, 1996.

[31] P. Betsch and E. Stein. An assumed strain approach avoiding artificial thickness straining for a non-linear 4-node shell element. *Communications in Numerical Methods in Engineering*, 11(11):899–909, 1995.

[32] J. E. Bischoff, E. M. Arruda, and K. Grosh. A microstructurally based orthotropic hyperelastic constitutive law. *Journal of Applied Mechanics*, 69(5):570–579, 2002.

[33] J. E. Bischoff, E. M. Arruda, and K. Grosh. Orthotropic hyperelasticity in terms of an arbitrary molecular chain model. *Journal of Applied Mechanics*, 69(2):198–201, 2002.

[34] M. Bischoff. *Theorie und Numerik einer dreidimensionalen Schalenformulierung*. PhD thesis, Institut für Baustatik, Universität Stuttgart, 1999.

[35] M. Bischoff and E. Ramm. Shear deformable shell elements for large strains and rotations. *International Journal for Numerical Methods in Engineering*, 40(23):4427–4449, 1997.

[36] M. Bischoff and I. Romero. A generalization of the method of incompatible modes. *International Journal for Numerical Methods in Engineering*, 69(9):1851–1868, 2007.

[37] M. Bischoff, W. A. Wall, K.-U. Bletzinger, and E. Ramm. *Encyclopedia of Computational Mechanics*, volume 2, chapter Models and Finite Elements for Thin-Walled Structures, pages 59–137. John Wiley & Sons, 2004.

[38] M. Böl. *Numerische Simulation von Polymernetzwerken mit Hilfe der Finite-Elemente-Methode*. PhD thesis, Institut für Mechanik, Ruhr-Universität Bochum, 2005.

[39] M. Böl and S. Reese. Finite element modelling of rubber-like polymers based on chain statistics. *International Journal of Solids and Structures*, 43(1):2–26, 2006.

[40] K.-U. Bletzinger, M. Bischoff, and E. Ramm. A unified approach for shear-locking-free triangular and rectangular shell finite elements. *Computers & Structures*, 75(3):321–334, 2000.

[41] P. Boisse, J. L. Daniel, and J. C. Gelin. A simple isoparametric three-node shell finite element. *Computers & Structures*, 44(6):1263–1273, 1992.

[42] J. Bonet. Large strain viscoelastic constitutive models. *International Journal of Solids and Structures*, 38(17):2953–2968, 2001.

[43] J. Bonet and A. J. Burton. A simple average nodal pressure tetrahedral element for incompressible and nearly incompressible dynamic explicit applications. *Communications In Numerical Methods In Engineering*, 14(5):437–449, 1998.

[44] J. Bonet, H. Marriott, and O. Hassan. Am averaged nodal deformation gradient linear tetrahedral element for large strain explicit dynamic applications. *Communications In Numerical Methods In Engineering*, 17(8):551–561, 2001.

[45] J. Bonet and R. D. Wood. *Nonlinear Continuum Mechanics for Finite Element Analysis*. Cambridge University Press, 2nd. edition, 2008.

[46] M. C. Boyce and E. M. Arruda. Constitutive models of rubber elasticity: A review. *Rubber chemistry and technology*, 73:504–523, 2000.

[47] B. Brank, J. Korelc, and A. Ibrahimbegovic. Nonlinear shell problem formulation accounting for through-the-thickness stretching and its finite element implementation. *Computers & Structures*, 2002.

[48] B. Brank, D. Peric, and F. B. Damjanic. On implementation of a nonlinear four node shell finite element for thin multilayered elastic shells. *Computational Mechanics*, 16(5):341–359, 1995.

[49] M. Breeuwer, S. de Putter, U. Kose, L. Speelman, K. Visser, F. Gerritsen, R. Hoogeveen, R. Krams, H. van den Bosch, J. Buth, T. Gunther, B. Wolters, E. van Dam, and F. van de Vosse. Towards patient-specific risk assessment of abdominal aortic aneurysm. *Medical and Biological Engineering and Computing*, 46(11):1085–1095, 2008.

[50] C. Bustamante, Z. Bryant, and S. B. Smith. Ten years of tension: single-molecule DNA mechanics. *Nature*, 421(6921):423–427, 2003.

[51] V. Calo, N. Brasher, Y. Bazilevs, and T. J. R. Hughes. Multiphysics model for blood flow and drug transport with application to patient-specific coronary artery flow. *Computational Mechanics*, 43(1):161–177, 2008.

[52] P. B. Canham, H. M. Finlay, J. G. Dixon, D. R. Boughner, and A. Chen. Measurements from light and polarized-light microscopy of human coronary-arteries fixed at distending pressure. *Cardiovascular Research*, 23(11):973–982, 1989.

[53] C. G. Caro, J. M. Fitzgerald, and R. C. Schroter. Atheroma and arterial wall shear — observation, correlation and proposal of a shear dependent mass transfer mechanism for altherogenesis. *Proceedings Of The Royal Society Of London Series B-Biological Sciences*, 177(1046):109–139, 1971.

[54] K. Chavan, B. Lamichhane, and B. Wohlmuth. Locking-free finite element methods for linear and nonlinear elasticity in 2d and 3d. *Computer Methods in Applied Mechanics and Engineering*, 196(41-44):4075–4086, 2007.

[55] C. J. Chuong and Y. C. Fung. Three-dimensional stress distribution in arteries. *Journal of Biomechanical Engineering*, 105:268–274, 1983.

[56] C. J. Chuong and Y. C. Fung. On residual stresses in arteries. *Journal of Biomechanical Engineering*, 108:189–192, 1986.

[57] A. Comerford and T. David. Computer model of nucleotide transport in a realistic porcine aortic trifurcation. *Annals of Biomedical Engineering*, 36(7):1175–1187, 2008.

[58] S. C. Cowin. Mechanical modeling of the stress-adaptation process in bone. *Calcified Tissue International*, 36:98–103, 1984.

[59] S. C. Cowin. Optimization of the strain energy density in linear anisotropic elasticity. *Journal of Elasticity*, 34(1):45–68, 1994.

[60] S. C. Cowin. How is a tissue built? *Journal of Biomechanical Engineering*, 122(6):553–569, 2000.

[61] S. C. Cowin and D. H. Hegedus. Bone remodeling I: theory of adaptive elasticity. *Journal Of Elasticity*, 6(3):313–326, 1976.

[62] M. Crisfield. *Non-linear Finite Element Analysis of Solids and Structures - Volume 2: Advanced Topics*. John Wiley & Sons, 1997.

[63] M. De Beule, P. Mortier, S. G. Carlier, B. Verhegghe, R. Van Impe, and P. Verdonck. Realistic finite element-based stent design: The impact of balloon folding. *Journal of Biomechanics*, 41(2):383–389, 2008.

[64] A. Delfino, N. Stergiopulos, J. E. Moore, and J. J. Meister. Residual strain effects on the stress field in a thick wall finite element model of the human carotid bifurcation. *Journal of Biomechanics*, 30(8):777–786, 1997.

[65] H. Demiray. A note on the elasticity of soft biological tissues. *Journal of Biomechanics*, 5(3):309–311, 1972.

[66] E. S. Di Martino, G. Guadagni, A. Fumero, G. Ballerini, R. Spirito, P. Biglioli, and A. Redaelli. Fluid-structure interaction within realistic three-dimensional models of the aneurysmatic aorta as a guidance to assess the risk of rupture of the aneurysm. *Medical Engineering & Physics*, 23(9):647–655, 2001.

[67] J. Djoko, B. Lamichhane, B. Reddy, and B. Wohlmuth. Conditions for equivalence between the hu-washizu and related formulations, and computational behavior in the incompressible limit. *Computer Methods in Applied Mechanics and Engineering*, 195(33-36):4161–4178, 2006.

[68] C. R. Dohrmann, M. W. Heinstein, J. Jung, S. W. Key, and W. R. Witkowski. Node-based uniform strain elements for three-node triangular and four-node tetrahedral meshes. *International Journal For Numerical Methods In Engineering*, 47(9):1549–1568, 2000.

[69] S. Doll. *Zur numerischen Behandlung großer elasto-viskoplastischer Deformationen bei isochor-volumetrisch entkoppeltem Stoffverhalten*. PhD thesis, Universität Karlsruhe (TH), 1998.

[70] S. Doll and K. Schweizerhof. On the development of volumetric strain energy functions. *Journal Of Applied Mechanics-Transactions Of The Asme*, 67(1):17–21, 2000.

[71] N. W. Dorland. *Dorland's Illustrated Medical Dictionary*. Saunders, 31 edition, 2007.

[72] N. J. B. Driessen, R. A. Boerboom, J. M. Huyghe, C. V. C. Bouten, and F. P. T. Baaijens. Computational analyses of mechanically induced collagen fiber remodeling in the aortic heart valve. *Journal of Biomechanical Engineering*, 125(4):549–557, 2003.

[73] N. J. B. Driessen, M. Cox, C. Bouten, and F. Baaijens. Remodelling of the angular collagen fiber distribution in cardiovascular tissues. *Biomechanics and Modeling in Mechanobiology*, 7(2):93–103, 2008.

[74] N. J. B. Driessen, G. W. M. Peters, J. M. Huyghe, C. V. C. Bouten, and F. P. T. Baaijens. Remodelling of continuously distributed collagen fibres in soft connective tissues. *Journal of Biomechanics*, 36(8):1151–1158, 2003.

[75] N. J. B. Driessen, W. Wilson, C. V. C. Bouten, and F. P. T. Baaijens. A computational model for collagen fibre remodelling in the arterial wall. *Journal of Theoretical Biology*, 226(1):53–64, 2004.

[76] A. Düster. *High order finite elements for three-dimensional, thin-walled nonlinear continua*. PhD thesis, Lehrstuhl für Bauinformatik, Technische Universität München, 2001.

[77] N. Duraiswamy, R. T. Schoephoerster, M. R. Moreno, and J. E. Moore. Stented artery flow patterns and their effects on the artery wall. *Annual Review of Fluid Mechanics*, 39(1):357–382, 2007.

[78] E. N. Dvorkin and K.-J. Bathe. A continuum mechanics based four-node shell element for general non-linear analysis. *Engineering Computations*, 1:77–88, 1984.

[79] F. Dyson. A meeting with Enrico Fermi. *Nature*, 427(6972):297–297, 2004.

[80] M. Epstein and G. A. Maugin. Thermomechanics of volumetric growth in uniform bodies. *International Journal Of Plasticity*, 16(7-8):951–978, 2000.

[81] S. Federico and W. Herzog. Towards an analytical model of soft biological tissues. *Journal of Biomechanics*, In Press, Corrected Proof, Available Online:–, 2008.

[82] C. A. Felippa. On the original publication of the general canonical functional of linear elasticity. *Journal of Applied Mechanics*, 67(1):217–219, 2000.

[83] C. A. Felippa. Supernatural QUAD4: a template formulation. *Computer Methods in Applied Mechanics and Engineering*, 195(41-43):5316–5342, 2006.

[84] C. A. Felippa and P. G. Bergan. A triangular bending element based on an energy-orthogonal free formulation. *Computer Methods in Applied Mechanics and Engineering*, 61(2):129–160, 1987.

[85] C. A. Figueroa, S. Baek, C. A. Taylor, and J. D. Humphrey. A computational framework for fluid-solid-growth modeling in cardiovascular simulations. *Computer Methods in Applied Mechanics and Engineering*, In Press, Accepted Manuscript, Available online:–, 2008.

[86] C. A. Figueroa, I. E. Vignon-Clementel, K. E. Jansen, T. J. R. Hughes, and C. A. Taylor. A coupled momentum method for modeling blood flow in three-dimensional deformable arteries. *Computer Methods in Applied Mechanics and Engineering*, 195(41-43):5685–5706, 2006.

[87] M. F. Fillinger, S. P. Marra, M. Raghavan, and F. E. Kennedy. Prediction of rupture risk in abdominal aortic aneurysm during observation: Wall stress versus diameter. *J Vasc Surg*, 37(4):724–732, 2003.

[88] M. F. Fillinger, M. Raghavan, S. P. Marra, J. L. Cronenwett, and F. E. Kennedy. In vivo analysis of mechanical wall stress and abdominal aortic aneurysm rupture risk. *Journal of Vascular Surgery*, 36(3):589–597, 2002.

[89] H. M. Finlay, L. McCullough, and P. B. Canham. Three-dimensional collagen organization of human brain arteries at different transmural pressures. *Journal Of Vascular Research*, 32(5):301–312, 1995.

[90] H. M. Finlay, P. Whittaker, and P. B. Canham. Collagen organization in the branching region of human brain arteries. *Stroke*, 29(8):1595–1601, 1998.

[91] D. P. Flanagan and T. Belytschko. A uniform strain hexahedron and quadrilateral with orthogonal hourglass control. *International Journal for Numerical Methods in Engineering*, 17(5):679–706, 1981.

[92] P. J. Flory. *Statistical Mechanics of Chain Molecules*. John Wiley & Sons, Inc., Chichester, New York, 1969.

[93] P. J. Flory and J. Rehner, Jr. Statistical mechanics of cross-linked polymer networks I. rubberlike elasticity. *Journal of Chemical Physics*, 11(11):512–520, 1943.

[94] L. Formaggia, J. F. Gerbeau, F. Nobile, and A. Quarteroni. On the coupling of 3d and 1d navier-stokes equations for flow problems in compliant vessels. *Computer Methods in Applied Mechanics and Engineering*, 191(6-7):561–582, 2001.

[95] P. Fratzl. *Collagen: Structure and Mechaics, an Introduction*, chapter 1, pages 1–13. Springer, 2008.

[96] P. Fratzl, editor. *Collagen, Structure and Mechanics*. Springer, 2008.

[97] M. A. Frenzel, M. Bischoff, K.-U. Bletzinger, and W. A. Wall. Performance of discrete strain gap (DSG) finite elements in the analysis of three-dimensional solids. In *Proceedings of the 5th International Conference on Computation of Shell and Spatial Structures*, Salzburg, Austria, 2005.

[98] M. A. Frenzel, M. Bischoff, K.-U. Bletzinger, and W. A. Wall. Performance of discrete strain gap (DSG) finite elements in the analysis of three-dimensional solids. In *1st GACM Colloquium on Computational Mechanics*, Ruhr University Bochum, Germany, 2005.

[99] M. A. Frenzel, M. Bischoff, and W. A. Wall. Solid finite elements for analysis of shells at large deformations, based on the DSG concept. In C. Mota Soares, J. A. C. Martins, H. C. Rodrigues, J. A. C. Ambrosio, C. A. B. Pina, C. M. Mota Soares, E. B. R. Pereira, and J. Folgado, editors, *III European Conference on Computational Mechanics - Solids, Structures and Coupled Problems in Engineering*, Lisbon, Portugal, 2006.

[100] M. A. Frenzel, C. Cyron, and W. A. Wall. Isogeometric structural shape optimization. In *2nd GACM Colloqium on Computational Mechanics*, Technische Universität München, Germany, 2007.

[101] Y. C. Fung. Biorheology of soft tissues. *Biorheology*, 10:139–155, 1973.

[102] Y. C. Fung. *Biomechanics: Mechanical properties of living tissues*. Springer, New York, 1983.

[103] Y. C. Fung. On the foundations of biomechanics. *Journal Of Applied Mechanics-Transactions Of The Asme*, 50(4B):1003–1009, 1983.

[104] Y. C. Fung. *Biodynamics: Circulation*. Springer, New York, 1984.

[105] Y. C. Fung. *Biomechanics: Motion, Flow, Stress, and Growth*. Springer, 1998.

[106] Y. C. Fung, K. Fronek, and P. Patitucci. Pseudoelasticity of arteries and the choice of its mathematical expression. *Am J Physiol Heart Circ Physiol*, 237(5):H620–631, 1979.

[107] K. Garikipati, E. M. Arruda, K. Grosh, H. Narayanan, and S. Calve. A continuum treatment of growth in biological tissue: the coupling of mass transport and mechanics. *Journal of the Mechanics and Physics of Solids*, 52(7):1595–1625, 2004.

[108] K. Garikipati, H. Narayanan, E. M. Arruda, K. Grosh, and S. Calve. Material forces in the context of biotissue remodelling. *Mechanics of Material Forces*, 11:77–84, 2005.

[109] K. Garikipati, J. Olberding, H. Narayanan, E. Arruda, K. Grosh, and S. Calve. Biological remodelling: Stationary energy, configurational change, internal variables and dissipation. *Journal of the Mechanics and Physics of Solids*, 54(7):1493–1515, 2006.

[110] T. C. Gasser, R. W. Ogden, and G. A. Holzapfel. Hyperelastic modelling of arterial layers with distributed collagen fibre orientations. *Journal of The Royal Society Interface*, 3(6):15–35, 2006.

[111] M. W. Gee, C. R. Dohrmann, S. W. Key, and W. A. Wall. A uniform nodal strain tetrahedron with isochoric stabilization. *International Journal for Numerical Methods in Engineering*, In Press:–, 2008.

[112] J.-F. Gerbeau and M. Vidrascu. A quasi-newton algorithm based on a reduced model for fluid-structure interaction problems in blood flows. *Esaim-Mathematical Modelling And Numerical Analysis-Modelisation Mathematique Et Analyse Numerique*, 37(4):631–647, 2003.

[113] J.-F. Gerbeau, M. Vidrascu, and P. Frey. Fluid-structure interaction in blood flows on geometries based on medical imaging. *Computers & Structures*, 83(2-3):155–165, 2005.

[114] R. Gleason and J. D. Humphrey. Effects of a sustained extension on arterial growth and remodeling: a theoretical study. *Journal of Biomechanics*, 38(6):1255–1261, 2005.

[115] S. Govindjee and J. C. Simo. Mullins effect and the strain amplitude dependence of the storage modulus. *International Journal of Solids and Structures*, 29(14-15):1737–1751, 1992.

[116] S. E. Greenwald, J. E. Moore, Jr., A. Rachev, T. P. C. Kane, and J.-J. Meister. Experimental investigation of the distribution of residual strains in the artery wall. *Journal of Biomechanical Engineering*, 119(4):438–444, 1997.

[117] F. Gruttmann, E. Stein, and P. Wriggers. Theory and numerics of thin elastic shells with finite rotations. *Archive of Applied Mechanics (Ingenieur Archiv)*, 59(1):54–67, 1989.

[118] A. Guillou and R. W. Ogden. *Mechanics of Biological Tissue*, chapter Growth in Soft Biological Tissue and Residual Stress Development, pages 47–62. Springer, Berlin, 2006.

[119] Y. Guo, M. Ortiz, T. Belytschko, and E. A. Repetto. Triangular composite finite elements. *International Journal For Numerical Methods In Engineering*, 47(1-3):287–316, 2000.

[120] H.-C. Han and Y.-C. Fung. Longitudinal strain of canine and porcine aortas. *Journal of Biomechanics*, 28(5):637–641, 1995.

[121] I. Hariton, G. deBotton, T. C. Gasser, and G. A. Holzapfel. Stress-driven collagen fiber remodeling in arterial walls. *Biomechanics and Modeling in Mechanobiology*, 6(3):163–175, 2007.

[122] I. Hariton, G. deBotton, T. C. Gasser, and G. A. Holzapfel. Stress-modulated collagen fiber remodeling in a human carotid bifurcation. *Journal of Theoretical Biology*, 248(3):460–470, 2007.

[123] M. Harnau. *Finite Volumen-Schalenelemente für große Deformationen und Kontakt*. PhD thesis, Institut für Baustatik, Universität Karlsruhe (TH), 2004.

[124] M. Harnau and K. Schweizerhof. About linear and quadratic 'solid-shell' elements at large deformations. *Computers & Structures*, 80(9-10):805–817, 2002.

[125] M. Harnau and K. Schweizerhof. Artificial kinematics and simple stabilization of solid-shell elements occurring in highly constrained situations and applications in composite sheet forming simulation. *Finite Elements in Analysis and Design*, 42(12):1097–1111, 2006.

[126] S. Hartmann. *Kontaktanalyse dünnwandiger Strukturen bei großen Deformationen*. PhD thesis, Institut für Baustatik und Baudynamik, Universität Stuttgart, 2007.

[127] H. W. Haslach. Nonlinear viscoelastic, thermodynamically consistent, models for biological soft tissue. *Biomechanics and Modeling in Mechanobiology*, 3(3):172–189, 2005.

[128] C. Haußer. *Effiziente Dreieckselemente für Flächentragwerke*. PhD thesis, Institut für Baustatik, Universität Stuttgart, 1996.

[129] C. Haußer and E. Ramm. Efficient 3-node shear deformable plate/shell elements - an almost hopeless undertaking. In B. H. V. Topping, editor, *Advances In Finite Element Technology*, pages 203–215, Budapest, 1996. Civil Comp Press.

[130] R. Hauptmann, S. Doll, M. Harnau, and K. Schweizerhof. 'solid-shell' elements with linear and quadratic shape functions at large deformations with nearly incompressible materials. *Computers & Structures*, 79(18):1671–1685, 2001.

[131] R. Hauptmann and K. Schweizerhof. A systematic development of 'solid-shell' element formulations for linear and non-linear analyses employing only displacement degrees of freedom. *International Journal for Numerical Methods in Engineering*, 42(1):49–69, 1998.

[132] R. Hauptmann, K. Schweizerhof, and S. Doll. Extension of the 'solid-shell' concept for application to large elastic and large elastoplastic deformations. *International Journal for Numerical Methods in Engineering*, 49(9):1121–1141, 2000.

[133] K. Hayashi, N. Stergiopulos, J.-J. Meister, S. E. Greenwald, and A. Rachev. *Cardiovascular techniques. Biomechanical systems: techniques and applications*, volume 2, chapter Techniques in the determination of the mechanical properties and constitutive laws of arterial walls. CRC Press, Boca Raton, 2001.

[134] M. Heil. An efficient solver for the fully coupled solution of large-displacement fluid-structure interaction problems. *Computer Methods in Applied Mechanics and Engineering*, 193(1-2):1–23, 2004.

[135] G. Himpel, A. Menzel, E. Kuhl, and P. Steinmann. Time-dependent fibre reorientation of transversely isotropic continua - finite element formulation and consistent linearization. *International Journal for Numerical Methods in Engineering*, 73(10):1413–1433, 2008.

[136] G. A. Holzapfel. On large strain viscoelasticity: continuum formulation and finite element applications to elastomeric structures. *International Journal for Numerical Methods in Engineering*, 39(22):3903–3926, 1996.

[137] G. A. Holzapfel. *Nonlinear solid mechanics - a continuum approach for engineering*. John Wiley & Sons, Inc., 2000.

[138] G. A. Holzapfel. *Encyclopedia of Computational Mechanics*, volume 2, chapter Computational Biomechanics of Soft Biological Tissue, pages 605–635. John Wiley & Sons, 2004.

[139] G. A. Holzapfel and T. C. Gasser. A viscoelastic model for fiber-reinforced composites at finite strains: Continuum basis, computational aspects and applications. *Computer Methods in Applied Mechanics and Engineering*, 190(34):4379–4403, 2001.

[140] G. A. Holzapfel, T. C. Gasser, and R. W. Ogden. A new constitutive framework for arterial wall mechanics and a comparative study of material models. *Journal of Elasticity*, 61(1):1–48, 2000.

[141] G. A. Holzapfel, T. C. Gasser, and M. Stadler. A structural model for the viscoelastic behavior of arterial walls: Continuum formulation and finite element analysis. *European Journal of Mechanics - A/Solids*, 21(3):441–463, 2002.

[142] G. A. Holzapfel, G. Sommer, M. Auer, P. Regitnig, and R. W. Ogden. Layer-specific 3d residual deformations of human aortas with non-atherosclerotic intimal thickening. *Annals of Biomedical Engineering*, 35(4):530–545, 2007.

[143] G. A. Holzapfel, G. Sommer, and P. Regitnig. Anisotropic mechanical properties of tissue components in human atherosclerotic plaques. *Journal of Biomechanical Engineering*, 126(5):657–665, 2004.

[144] G. A. Holzapfel, M. Stadler, and T. C. Gasser. Changes in the mechanical environment of stenotic arteries during interaction with stents: Computational assessment of parametric stent designs. *Journal Of Biomechanical Engineering-Transactions Of The Asme*, 127(1):166–180, 2005.

[145] G. A. Holzapfel, M. Stadler, and C. A. J. Schulze-Bauer. A layer-specific three-dimensional model for the simulation of balloon angioplasty using magnetic resonance imaging and mechanical testing. *Annals of Biomedical Engineering*, 30(6):753–767, 2002.

[146] G. A. Holzapfel and H. W. Weizsäcker. Biomechanical behavior of the arterial wall and its numerical characterization. *Computers in Biology and Medicine*, 28(4):377–392, 1998.

[147] T. J. R. Hughes. *The Finite Element Method–Linear Static and Dynamic Finite Element Analysis*. Dover Publications, New York, 1987.

[148] T. J. R. Hughes, J. A. Cottrell, and Y. Bazilevs. Isogeometric analysis: CAD, finite elements, NURBS, exact geometry and mesh refinement. *Computer Methods in Applied Mechanics and Engineering*, 194(39-41):4135–4195, 2005.

[149] T. J. R. Hughes and R. L. Taylor. *The mathematics of finite elements and applications IV*, chapter The linear triangular bending element, pages 127–142. Academic Press, New York, 1981.

[150] T. J. R. Hughes, R. L. Taylor, and W. Kanoknukulchai. A simple and efficient finite element for plate bending. *International Journal for Numerical Methods in Engineering*, 11(10):1529–1543, 1977.

[151] T. J. R. Hughes and T. Tezduyar. Finite elements based upon Mindlin plate theory with particular reference to the four-node isoparametric element. *Journal of Applied Mechanics*, 48:265–284, 1981.

[152] J. D. Humphrey. *Cardiovascular Solid Mechanics: Cells, Tissues and Organs*. Springer, New York, 2002.

[153] J. D. Humphrey, J. Eberth, W. Dye, and R. Gleason. Fundamental role of axial stress in compensatory adaptations by arteries. *Journal of Biomechanics*, 42(1):1–8, 2009.

[154] J. D. Humphrey and S. Na. Elastodynamics and arterial wall stress. *Annals of Biomedical Engineering*, 30(4):509–523, 2002.

[155] J. D. Humphrey and K. R. Rajagopal. A constrained mixture model for growth and remodeling of soft tissues. *Mathematical Models & Methods In Applied Sciences*, 12(3):407–430, 2002.

[156] J. D. Humphrey and C. Taylor. Intracranial and abdominal aortic aneurysms: Similarities, differences, and need for a new class of computational models. *Annual Review of Biomedical Engineering*, 10(1):221–246, 2008.

[157] B. Irons and S. Ahmad. *Techniques of finite elements*. E. Horwood ; Halsted Press, Chichester, 1980.

[158] B. M. Irons. Numerical integration applied to finite element methods. In *Conference on Use of Digital Computers in Structural Engineering*. University of Newcastle, 1966.

[159] M. Itskov. *Tensor Algebra and Tensor Analysis for Engineers*. Springer, New York, 2007.

[160] M. Itskov, A. Ehret, and D. Mavrilas. A polyconvex anisotropic strain–energy function for soft collagenous tissues. *Biomechanics and Modeling in Mechanobiology*, 5(1):17–26, 2006.

[161] M. Kaazempur-Mofrad and C. Ethier. Mass transport in an anatomically realistic human right coronary artery. *Annals of Biomedical Engineering*, 29(2):121–127, 2001.

[162] M. R. Kaazempur-Mofrad, A. G. Isasi, H. F. Younis, R. C. Chan, D. P. Hinton, G. Sukhova, G. M. LaMuraglia, R. T. Lee, and R. D. Kamm. Characterization of the atherosclerotic carotid bifurcation using MRI, finite element modeling, and histology. *Annals of Biomedical Engineering*, 32(7):932–946, 2004.

[163] M. Kaliske. A formulation of elasticity and viscoelasticity for fibre reinforced material at small and finite strains. *Computer Methods in Applied Mechanics and Engineering*, 185(2-4):225–243, 2000.

[164] P. Kalita and R. Schaefer. Mechanical models of artery walls. *Archives of Computational Methods in Engineering*, 15(1):1–36, 2008.

[165] E. P. Kasper and R. L. Taylor. A mixed-enhanced strain method: Part I: geometrically linear problems. *Computers & Structures*, 75(3):237–250, 2000.

[166] E. P. Kasper and R. L. Taylor. A mixed-enhanced strain method: Part II: geometrically nonlinear problems. *Computers & Structures*, 75(3):251–260, 2000.

[167] S. W. Key, A. S. Gullerud, and J. R. Koteras. A low-order, hexahedral finite element for modelling shells. *International Journal for Numerical Methods in Engineering*, 59(7):923–944, 2004.

[168] C. Kleinstreuer, Z. Li, and M. A. Farber. Fluid-structure interaction analyses of stented abdominal aortic aneurysms. *Annual Review of Biomedical Engineering*, 9(1):169–204, 2007.

[169] S. Klinkel. *Theorie und Numerik eines Volumen-Schalen-Elementes bei finiten elastischen und plastischen Verzerrungen*. PhD thesis, Institut für Baustatik, Universität Karlsruhe (TH), 2000.

[170] S. Klinkel, F. Gruttmann, and W. Wagner. A continuum based three-dimensional shell element for laminated structures. *Computers & Structures*, 1999.

[171] S. Klinkel, F. Gruttmann, and W. Wagner. A robust non-linear solid shell element based on a mixed variational formulation. *Computer Methods in Applied Mechanics and Engineering*, 195(1-3):179–201, 2006.

[172] S. Klinkel, F. Gruttmann, and W. Wagner. A mixed shell formulation accounting for thickness strains and finite strain 3d material models. *International Journal for Numerical Methods in Engineering*, 74(6):945–970, 2008.

[173] S. Klinkel and W. Wagner. A geometrical non-linear brick element based on the EAS-method. *International Journal for Numerical Methods in Engineering*, 40(24):4529–4545, 1997.

[174] S. Kondo, N. Hashimoto, H. Kikuchi, F. Hazama, I. Nagata, H. Kataoka, and R. L. Macdonald. Cerebral aneurysms arising at nonbranching sites: An experimental study. *Stroke*, 28(2):398–404, 1997.

[175] F. Koschnick. *Geometrische Locking-Effekte bei Finiten Elementen und ein allgemeines Konzept zu ihrer Vermeidung*. PhD thesis, Lehrstuhl für Statik, Technische Universität München, 2004.

[176] F. Koschnick, M. Bischoff, N. Camprubí, and K.-U. Bletzinger. The discrete strain gap method and membrane locking. *Computer Methods in Applied Mechanics and Engineering*, 194(21-24):2444–2463, 2005.

[177] O. Kratky and G. Porod. Röntgenuntersuchung Gelöster Fadenmoleküle. *Recueil des Travaux Chimiques des Pays-Bas-Journal of the Royal Netherlands Chemical Society*, 68(12):1106–1122, 1949.

[178] M. Kroon and G. A. Holzapfel. A model for saccular cerebral aneurysm growth by collagen fibre remodelling. *Journal Of Theoretical Biology*, 247(4):775–787, 2007.

[179] M. Kroon and G. A. Holzapfel. A theoretical model for saccular cerebral aneurysm growth: Deformation and stress-analysis. *Proceeding Of The Amse Summer Bioengineering Conference - 2007*, pages 255–256, 2007.

[180] M. Kroon and G. A. Holzapfel. Modeling of saccular aneurysm growth in a human middle cerebral artery. *Journal Of Biomechanical Engineering-Transactions Of The Asme*, 130(5):051012, 2008.

[181] M. Kroon and G. A. Holzapfel. A new constitutive model for multi-layered collagenous tissues. *Journal of Biomechanics*, 41(12):2766–2771, 2008.

[182] M. Kroon and G. A. Holzapfel. A theoretical model for fibroblast-controlled growth of saccular cerebral aneurysms. *Journal of Theoretical Biology*, 257(1):73–83, 2009.

[183] U. Küttler, M. Gee, C. Förster, A. Comerford, and W. A. Wall. Coupling strategies for biomedical fluid-structure interaction problems. *Special Issue on "Fluid-Structure Interaction in Biomedical Applications" of Communications in Numerical Methods in Engineering*, submitted:–, 2008.

[184] E. Kuhl, K. Garikipati, E. M. Arruda, and K. Grosh. Remodeling of biological tissue: Mechanically induced reorientation of a transversely isotropic chain network. *Journal of the Mechanics and Physics of Solids*, 53(7):1552–1573, 2005.

[185] E. Kuhl and G. A. Holzapfel. A continuum model for remodeling in living structures. *Journal of Materials Science*, 42(21):8811–8823, 2007.

[186] E. Kuhl, R. Maas, G. Himpel, and A. Menzel. Computational modeling of arterial wall growth. *Biomechanics and Modeling in Mechanobiology*, 6(5):321–331, 2007.

[187] E. Kuhl and P. Steinmann. Theory and numerics of geometrically non-linear open system mechanics. *International Journal For Numerical Methods In Engineering*, 58(11):1593–1615, 2003.

[188] W. Kuhn. Beziehungen zwischen Molekülgröße, statistischer Molekülgestalt und elastischen Eigenschaften hochpolymerer Stoffe. *Colloid & Polymer Science*, 76(3):258–271, 1936.

[189] W. Kuhn and F. Grün. Beziehungen zwischen elastischen konstanten und dehnungsdoppelbrechung hochelastischer stoffe. *Colloid & Polymer Science*, 101(3):248–271, 1942.

[190] Y. Lanir. Constitutive equations for fibrous connective tissues. *Journal of Biomechanics*, 16(1):1–12, 1983.

[191] P. Le Tallec and J. Mouro. Fluid structure interaction with large structural displacements. *Computer Methods in Applied Mechanics and Engineering*, 190(24-25):3039–3067, 2001.

[192] A. Legay and A. Combescure. Elastoplastic stability analysis of shells using the physically stabilized finite element. *International Journal for Numerical Methods in Engineering*, 57(9):1299–1322, 2003.

[193] M. Lei, J. P. Archie, and C. Kleinstreuer. Computational design of a bypass graft that minimizes wall shear stress gradients in the region of the distal anastomosis. *Journal of Vascular Surgery*, 25(4):637–646, 1997.

[194] A. Leuprecht, S. Kozerke, P. Boesiger, and K. Perktold. Blood flow in the human ascending aorta: a combined MRI and CFD study. *Journal of Engineering Mathematics*, 47(3):387–404, 2003.

[195] A. Leuprecht, K. Perktold, M. Prosi, T. Berk, W. Trubel, and H. Schima. Numerical study of hemodynamics and wall mechanics in distal end-to-side anastomoses of bypass grafts. *Journal of Biomechanics*, 35(2):225–236, 2002.

[196] Z. Li and C. Kleinstreuer. Blood flow and structure interactions in a stented abdominal aortic aneurysm model. *Medical Engineering & Physics*, 27(5):369–382, 2005.

[197] S. Q. Liu and Y. C. Fung. Zero-stress states of arteries. *Journal Of Biomechanical Engineering-Transactions Of The Asme*, 110(1):82–84, 1988.

[198] S. Q. Liu and Y. C. Fung. Influence of STZ-induced diabetes on zero-stress states of rat pulmonary and systemic arteries. *Diabetes*, 41(2):136–146, 1992.

[199] V. A. Lubarda. Constitutive theories based on the multiplicative decomposition of deformation gradient: Thermoelasticity, elastoplasticity, and biomechanics. *Applied Mechanics Reviews*, 57(2):95–108, 2004.

[200] V. A. Lubarda and A. Hoger. On the mechanics of solids with a growing mass. *International Journal of Solids and Structures*, 39(18):4627–4664, 2002.

[201] J. Lubliner. A model of rubber viscoelasticity. *Mechanics Research Communications*, 12(2):93–99, 1985.

[202] R. H. Macneal. A simple quadrilateral shell element. *Computers & Structures*, 8(2):175–183, 1978.

[203] R. H. MacNeal. A theorem regarding the locking of tapered four-noded membrane elements. *International Journal for Numerical Methods in Engineering*, 24:1793–1799, 1987.

[204] R. H. MacNeal. On the limits of finite element perfectability. *International Journal for Numerical Methods in Engineering*, 35(8):1589–1601, 1992.

[205] R. H. MacNeal. *Finite Elements: Their Design an Performance*. Dekker, 1994.

[206] R. H. Macneal and R. L. Harder. A proposed standard set of problems to test finite element accuracy. *Finite Elements in Analysis and Design*, 1(1):3–20, 1985.

[207] J. F. Marko and E. D. Siggia. Stretching DNA. *Macromolecules*, 28(26):8759–8770, 1995.

[208] J. E. Marsden and T. J. R. Hughes. *Mathematical Foundations of Elasticity*. Prentice-Hall, Englewood Cliffs, New Jersey, 1983.

[209] T. Matsumoto and K. Hayashi. Stress and strain distribution in hypertensive and normotensive rat aorta considering residual strain. *Journal Of Biomechanical Engineering-Transactions Of The Asme*, 118(1):62–73, 1996.

[210] D. A. McDonald. *Blood Flow in Arteries*. Williams & Wilkins, Baltimore, 2nd edition, 1974.

[211] H. Meng, Z. Wang, Y. Hoi, L. Gao, E. Metaxa, D. D. Swartz, and J. Kolega. Complex hemodynamics at the apex of an arterial bifurcation induces vascular remodeling resembling cerebral aneurysm initiation. *Stroke*, 38(6):1924–1931, 2007.

[212] A. Menzel. Modelling of anisotropic growth in biological tissues. *Biomechanics and Modeling in Mechanobiology*, 3(3):147–171, 2005.

[213] A. Menzel. A fibre reorientation model for orthotropic multiplicative growth. *Biomechanics and Modeling in Mechanobiology*, 6(5):303–320, 2007.

[214] C. Miehe. A theoretical and computational model for isotropic elastoplastic stress analysis in shells at large strains. *Computer Methods in Applied Mechanics and Engineering*, 155(3-4):193–233, 1998.

[215] C. Miehe, S. Göktepe, and F. Lulei. A micro-macro approach to rubber-like materials–part I: the non-affine micro-sphere model of rubber elasticity. *Journal of the Mechanics and Physics of Solids*, 52(11):2617–2660, 2004.

[216] F. Migliavacca, F. Gervaso, M. Prosi, P. Zunino, S. Minisini, L. Formaggia, and G. Dubini. Expansion and drug elution model of a coronary stent. *Computer Methods in Biomechanics and Biomedical Engineering*, 10(1):63–73, 2007.

[217] K. Nakshatrala, A. Masud, and K. Hjelmstad. On finite element formulations for nearly incompressible linear elasticity. *Computational Mechanics*, 41(4):547–561, 2008.

[218] B. Nedjar. An anisotropic viscoelastic fibre-matrix model at finite strains: Continuum formulation and computational aspects. *Computer Methods in Applied Mechanics and Engineering*, 196(9-12):1745–1756, 2007.

[219] E. A. S. Neto, F. M. A. Pires, and D. R. J. Owen. F-bar-based linear triangles and tetrahedra for finite strain analysis of nearly incompressible solids. Part I: formulation and benchmarking. *International Journal for Numerical Methods in Engineering*, 62(3):353–383, 2005.

[220] R. W. Ogden. *Non-Linear Elastic Deformations*. Ellis Horwood, Chichester, 1984.

[221] R. W. Ogden. *Biomechanics of soft tissue in cardiovascular systems*, chapter Nonlinear elasticity, anisotropy, material stability and residual stresses in soft tissue, pages 65–108. Springer, New York, 2003.

[222] J. Ohayon, P. Teppaz, G. Finet, and G. Rioufol. In-vivo prediction of human coronary plaque rupture location using intravascular ultrasound and the finite element method. *Coronary Artery Disease*, 12:655–663, 2001.

[223] H. Parisch. A continuum-based shell theory for non-linear applications. *International Journal for Numerical Methods in Engineering*, 38(11):1855–1883, 1995.

[224] H. Parisch. *Festkörper-Kontinuumsmechanik*. Teubner Verlag, Stuttgart, 2003.

[225] H. C. Park, C. Cho, and S. W. Lee. An efficient assumed strain element model with six DOF per node for geometrically non-linear shells. *International Journal for Numerical Methods in Engineering*, 38(24):4101–4122, 1995.

[226] K. C. Park and G. M. Stanley. A curved C0-shell element based on assumed natural-coordinate strains. *Journal of Applied Mechanics*, 53:278–290, 1986.

[227] E. Peña, B. Calvo, M. A. Martínez, and M. Doblaré. On finite-strain damage of viscoelastic-fibred materials. Application to soft biological tissues. *International Journal for Numerical Methods in Engineering*, 74(7):1198–1218, 2008.

[228] X. Peng and M. A. Crisfield. A consistent co-rotational formulation for shells using the constant stress/constant moment triangle. *International Journal for Numerical Methods in Engineering*, 35(9):1829–1847, 1992.

[229] K. Perktold, H. Florian, and D. Hilbert. Analysis of pulsatile blood flow: a carotid siphon model. *Journal of Biomedical Engineering*, 9:46–53, 1987.

[230] K. Perktold and D. Hilbert. Numerical simulation of pulsatile flow in a carotid bifurcation model. *Journal of Biomedical Engineering*, 8:193–199, 1986.

[231] K. Perktold and G. Rappitsch. Computer simulation of local blood flow and vessel mechanics in a compliant carotid artery bifurcation model. *Journal of Biomechanics*, 28(7):845–856, 1995.

[232] T. H. H. Pian and D.-P. Chen. Alternative ways for formulation of hybrid stress elements. *International Journal for Numerical Methods in Engineering*, 18(11):1679–1684, 1982.

[233] T. H. H. Pian and K. Sumihara. Rational approach for assumed stress finite elements. *International Journal for Numerical Methods in Engineering*, 20(9):1685–1695, 1984.

[234] T. H. H. Pian and P. Tong. Relations between incompatible displacement model and hybrid stress model. *International Journal for Numerical Methods in Engineering*, 22(1):173–181, 1986.

[235] F. M. A. Pires, E. A. D. Neto, and J. L. D. Padilla. An assessment of the average nodal volume formulation for the analysis of nearly incompressible solids under finite strains. *Communications In Numerical Methods In Engineering*, 20(7):569–583, 2004.

[236] M. A. Puso and J. Solberg. A stabilized nodally integrated tetrahedral. *International Journal For Numerical Methods In Engineering*, 67(6):841–867, 2006.

[237] A. Quarteroni, M. Tuveri, and A. Veneziani. Computational vascular fluid dynamics: problems, models and methods. *Computing and Visualization in Science*, 2(4):163–197, 2000.

[238] A. Rachev. A model of arterial adaptation to alterations in blood flow. *Journal of Elasticity*, 61(1):83–111, 2000.

[239] A. Rachev and K. Hayashi. Theoretical study of the effects of vascular smooth muscle contraction on strain and stress distributions in arteries. *Annals of Biomedical Engineering*, 27(4):459–468, 1999.

[240] A. Rachev, N. Stergiopulos, and J.-J. Meister. A model for geometric and mechanical adaptation of arteries to sustained hypertension. *Journal of Biomechanical Engineering*, 120(1):9–17, 1998.

[241] E. Ramm. Form und Tragverhalten. In E. Ramm and E. Schunck, editors, *Heinz Isler Schalen*. Krämer, Stuttgart, 1986.

[242] E. Ramm, M. Bischoff, and M. Braun. Higher order nonlinear shell formulation - a step back into three dimensions. In K. Bell, editor, *From Finite Elements to the Troll Platform, Ivar Holland 70th Anniversary*, pages 65–88. The Norwegian Institute of Technology, Trondheim, 1994.

[243] S. Reese. A large deformation solid-shell concept based on reduced integration with hourglass stabilization. *International Journal for Numerical Methods in Engineering*, 69(8):1671–1716, 2007.

[244] S. Reese and S. Govindjee. A theory of finite viscoelasticity and numerical aspects. *International Journal of Solids and Structures*, 35(26-27):3455–3482, 1998.

[245] S. Reese and C. Rickelt. A model-adaptive hanging node concept based on a new nonlinear solid-shell formulation. *Computer Methods in Applied Mechanics and Engineering*, 197(1-4):61–79, 2007.

[246] J. A. G. Rhodin. *Handbook of physiology, the cardiovascular system*, volume 2, chapter Architecture of the vessel wall, pages 1–31. American Physiological Society, Bethesda, Maryland, 1980.

[247] P. D. Richardson. Biomechanics of plaque rupture: Progress, problems, and new frontiers. *Annals of Biomedical Engineering*, 30(4):524–536, 2002.

[248] M. R. Roach and A. C. Burton. The reason for the shape of distensibility curve of arteries. *Canadian journal of biochemistry and physiology*, 35(8):681–690, 1957.

[249] J. F. Rodríguez, V. Alastrué, and M. Doblaré. Finite element implementation of a stochastic three dimensional finite-strain damage model for fibrous soft tissue. *Computer Methods in Applied Mechanics and Engineering*, 197(9-12):946–958, 2008.

[250] J. F. Rodríguez, F. Cacho, J. A. Bea, and M. Doblaré. A stochastic-structurally based three dimensional finite-strain damage model for fibrous soft tissue. *Journal of the Mechanics and Physics of Solids*, 54(4):864–886, 2006.

[251] J. F. Rodríguez, C. Ruiz, M. Doblaré, and G. A. Holzapfel. Mechanical stresses in abdominal aortic aneurysms: Influence of diameter, asymmetry, and material anisotropy. *Journal of Biomechanical Engineering*, 130(2):021023–10, 2008.

[252] E. K. Rodriguez, A. Hoger, and A. D. McCulloch. Stress-dependent finite growth in soft elastic tissues. *Journal of Biomechanics*, 27(4):455–467, 1994.

[253] A. J. Rowe, H. M. Finlay, and P. B. Canham. Collagen biomechanics in cerebral arteries and bifurcations assessed by polarizing microscopy. *Journal of Vascular Research*, 40:406–414, 2003.

[254] O. Sahni, K. Jansen, C. Taylor, and M. Shephard. Automated adaptive cardiovascular flow simulations. *Engineering with Computers*, 25(1):25–36, 2009.

[255] C. Sansour and F. G. Kollmann. Families of 4-node and 9-node finite elements for a finite deformation shell theory. An assessment of hybrid stress, hybrid strain and enhanced strain elements. *Computational Mechanics*, 24(6):435–447, 2000.

[256] R. Schleebusch. *Theorie und Numerik einer oberflächenorientierten Schalenformulierung*. PhD thesis, Institut für Mechanik und Flächentragwerke, Universität Dresden, 2005.

[257] H. Schoop. Oberflächenorientierte Schalentheorien endlicher Verschiebungen. *Ingenieur-Archiv*, 56:427–437, 1986.

[258] J. Schröder and P. Neff. On the construction of polyconvex anisotropic free energy functions. In C. Miehe, editor, *Proceedings of the IUTAM Symposium on Computational Mechanics of Solid Materials at Large Strains*, page 171–180, Dordrecht, 2001. Kluwer Academic Publishers.

[259] J. Schröder and P. Neff. Invariant formulation of hyperelastic transverse isotropy based on polyconvex free energy functions. *International Journal of Solids and Structures*, 40(2):401–445, 2003.

[260] J. Schröder, P. Neff, and D. Balzani. A variational approach for materially stable anisotropic hyperelasticity. *International Journal of Solids and Structures*, 42(15):4352–4371, 2005.

[261] C. Scotti, A. Shkolnik, S. Muluk, and E. Finol. Fluid-structure interaction in abdominal aortic aneurysms: effects of asymmetry and wall thickness. *BioMedical Engineering OnLine*, 4(1):64, 2005.

[262] R. R. Seeley, T. D. Stephens, and P. Tate. *Anatomy and physiology*. McGraw-Hill, Boston, 6th edition, 2003.

[263] J. C. Simo. On a fully three-dimensional finite-strain viscoelastic damage model: Formulation and computational aspects. *Computer Methods in Applied Mechanics and Engineering*, 60(2):153–173, 1987.

[264] J. C. Simo. A framework for finite strain elastoplasticity based on maximum plastic dissipation and the multiplicative decomposition: Part I: continuum formulation. *Computer Methods in Applied Mechanics and Engineering*, 66(2):199–219, 1988.

[265] J. C. Simo. A framework for finite strain elastoplasticity based on maximum plastic dissipation and the multiplicative decomposition: Part II: computational aspects. *Computer Methods in Applied Mechanics and Engineering*, 68(1):1–31, 1988.

[266] J. C. Simo. Algorithms for static and dynamic multiplicative plasticity that preserve the classical return mapping schemes of the infinitesimal theory. *Computer Methods in Applied Mechanics and Engineering*, 99(1):61–112, 1992.

[267] J. C. Simo and F. Armero. Geometrically non-linear enhanced strain mixed methods and the method of incompatible modes. *International Journal for Numerical Methods in Engineering*, 33(7):1413–1449, 1992.

[268] J. C. Simo, F. Armero, and R. L. Taylor. Improved versions of assumed enhanced strain tri-linear elements for 3d finite deformation problems. *Computer Methods in Applied Mechanics and Engineering*, 110(3-4):359–386, 1993.

[269] J. C. Simo and T. J. R. Hughes. On variational foundations of assumed strain methods. *Journal of Applied Mechanics*, 53:51–54, 1986.

[270] J. C. Simo and T. J. R. Hughes. *Computational Inelasticity*. Springer, New York, 1998.

[271] J. C. Simo and M. Ortiz. A unified approach to finite deformation elastoplastic analysis based on the use of hyperelastic constitutive equations. *Computer Methods in Applied Mechanics and Engineering*, 49(2):221–245, 1985.

[272] J. C. Simo and M. S. Rifai. A class of mixed assumed strain methods and the method of incompatible modes. *International Journal for Numerical Methods in Engineering*, 29(8):1595–1638, 1990.

[273] J. C. Simo, M. S. Rifai, and D. D. Fox. On a stress resultant geometrically exact shell model. Part IV: variable thickness shells with through-the-thickness stretching. *Computer Methods in Applied Mechanics and Engineering*, 81(1):91–126, 1990.

[274] J. C. Simo and R. L. Taylor. Quasi-incompressible finite elasticity in principal stretches. continuum basis and numerical algorithms. *Computer Methods in Applied Mechanics and Engineering*, 85(3):273–310, 1991.

[275] J. C. Simo, R. L. Taylor, and K. S. Pister. Variational and projection methods for the volume constraint in finite deformation elasto-plasticity. *Computer Methods in Applied Mechanics and Engineering*, 51(1-3):177–208, 1985.

[276] R. Skalak, G. Dasgupta, M. Moss, E. Otten, P. Dullemeijer, and H. Vilmann. Analytical description of growth. *Journal of Theoretical Biology*, 94(3):555–577, 1982.

[277] A. J. M. Spencer. *Continuum Physics*, chapter Theory of invariants, pages 239–253. Academic Press, 1954.

[278] D. Steinman, D. Vorp, and C. Ethier. Computational modeling of arterial biomechanics: Insights into pathogenesis and treatment of vascular disease. *Journal of Vascular Surgery*, 37(5):1118–1128, 2003.

[279] D. A. Steinman. Image-based computational fluid dynamics modeling in realistic arterial geometries. *Annals of Biomedical Engineering*, 30(4):483–497, 2002.

[280] D. A. Steinman, J. B. Thomas, H. M. Ladak, J. S. Milner, B. K. Rutt, and J. D. Spence. Reconstruction of carotid bifurcation hemodynamics and wall thickness using computational fluid dynamics and MRI. *Magnetic Resonance in Medicine*, 47(1):149–159, 2002.

[281] K. Y. Sze. On immunizing five-beta hybrid-stress element models from 'trapezoidal locking' in practical analyses. *International Journal for Numerical Methods in Engineering*, 47(4):907–920, 2000.

[282] K. Y. Sze and W. K. Chan. A six-node pentagonal assumed natural strain solid-shell element. *Finite Elements in Analysis and Design*, 37(8):639–655, 2001.

[283] K. Y. Sze, W. K. Chan, and T. H. H. Pian. An eight-node hybrid-stress solid-shell element for geometric non-linear analysis of elastic shells. *International Journal for Numerical Methods in Engineering*, 55(7):853–878, 2002.

[284] K. Y. Sze, X. H. Liu, and S. H. Lo. Hybrid-stress six-node prismatic elements. *International Journal for Numerical Methods in Engineering*, 61(9):1451–1470, 2004.

[285] K. Y. Sze, X. H. Liu, and S. H. Lo. Popular benchmark problems for geometric nonlinear analysis of shells. *Finite Elements in Analysis and Design*, 40(11):1551–1569, 2004.

[286] K. Y. Sze and L. Q. Yao. A hybrid stress ANS solid-shell element and its generalization for smart structure modelling. Part I — solid-shell element formulation. *International Journal for Numerical Methods in Engineering*, 48(4):545–564, 2000.

[287] K. Y. Sze, S. J. Zheng, and S. H. Lo. A stabilized eighteen-node solid element for hyperelastic analysis of shells. *Finite Elements in Analysis and Design*, 40(3):319–340, 2004.

[288] L. A. Taber. Biomechanics of growth, remodelling, and morphogenesis. *Applied Mechanics Reviews*, 48:487–545, 1995.

[289] L. A. Taber. A model for aortic growth based on fluid shear and fiber stresses. *J. Biomech. Eng.*, 120(3):348–354, 1998.

[290] L. A. Taber and J. D. Humphrey. Stress-modulated growth, residual stress, and vascular heterogeneity. *Journal Of Biomechanical Engineering-Transactions Of The Asme*, 123(6):528–535, 2001.

[291] K. Takamizawa and K. Hayashi. Strain energy density function and uniform strain hypothesis for arterial mechanics. *Journal of Biomechanics*, 20(1):7–17, 1987.

[292] X. G. Tan and L. Vu-Quoc. Efficient and accurate multilayer solid-shell element: nonlinear materials at finite strain. *International Journal for Numerical Methods in Engineering*, 63(15):2124–2170, 2005.

[293] C. A. Taylor. *Encyclopedia of Computational Mechanics*, volume 3, chapter Blood Flow, pages 527–543. John Wiley & Sons, 2004.

[294] C. A. Taylor and M. T. Draney. Experimental and computational methods in cardiovascular fluid mechanics. *Annual Review of Fluid Mechanics*, 36(1):197–231, 2004.

[295] C. A. Taylor, M. T. Draney, J. P. Ku, D. Parker, B. N. Steele, K. Wang, and C. K. Zarins. Predictive medicine: Computational techniques in therapeutic decision-making. *Computer Aided Surgery*, 4(5):231–247, 1999.

[296] C. A. Taylor, T. J. R. Hughes, and C. K. Zarins. Finite element modeling of blood flow in arteries. *Computer Methods in Applied Mechanics and Engineering*, 158(1-2):155–196, 1998.

[297] C. A. Taylor, T. J. R. Hughes, and C. K. Zarins. Finite element modeling of three-dimensional pulsatile flow in the abdominal aorta: Relevance to atherosclerosis. *Annals of Biomedical Engineering*, 26(6):975–987, 1998.

[298] R. L. Taylor. A mixed-enhanced formulation tetrahedral finite elements. *International Journal for Numerical Methods in Engineering*, 47(1-3):205–227, 2000.

[299] R. L. Taylor, P. J. Beresford, and E. L. Wilson. A non-conforming element for stress analysis. *International Journal for Numerical Methods in Engineering*, 10(6):1211–1219, 1976.

[300] P. Thoutireddy, J. F. Molinari, E. A. Repetto, and M. Ortiz. Tetrahedral composite finite elements. *International Journal for Numerical Methods in Engineering*, 53(6):1337–1351, 2002.

[301] M. J. Thubrikar. *Vascular Mechanics and Pathology*. Springer, 2007.

[302] M. J. Thubrikar, J. Al-Soudi, and F. Robicsek. Wall stress studies of abdominal aortic aneurysm in a clinical model. *Annals of Vascular Surgery*, 15(3):355–366, 2001.

[303] M. J. Thubrikar, M. Labrosse, F. Robicsek, J. Al-Soudi, and B. Fowler. Mechanical properties of abdominal aortic aneurysm wall. *Journal of Medical Engineering & Technology*, 25(4):133–142, 2001.

[304] M. J. Thubrikar and F. Robicsek. Pressure-induced arterial wall stress and atherosclerosis. *The Annals of Thoracic Surgery*, 59(6):1594–1603, 1995.

[305] E. Tonti. *On the formal structure of physical theories*. Monograph of the Italian National Research Council, 1975.

[306] R. Torii, M. Oshima, T. Kobayashi, K. Takagi, and T. Tezduyar. Fluid–structure interaction modeling of a patient-specific cerebral aneurysm: influence of structural modeling. *Computational Mechanics*, 43(1):151–159, 2008.

[307] R. Torii, M. Oshima, T. Kobayashi, K. Takagi, and T. E. Tezduyar. Computer modeling of cardiovascular fluid-structure interactions with the deforming-spatial-domain/stabilized space-time formulation. *Computer Methods in Applied Mechanics and Engineering*, 195(13-16):1885–1895, 2006.

[308] R. Torii, M. Oshima, T. Kobayashi, K. Takagi, and T. E. Tezduyar. Fluid-structure interaction modeling of blood flow and cerebral aneurysm: Significance of artery and aneurysm shapes. *Computer Methods in Applied Mechanics and Engineering*, In Press, Corrected Proof:–, 2008.

[309] L. R. G. Treloar. The elasticity of a network of long-chain molecules I/II. *Transactions of the Faraday Society*, 39:36–41, 241–246, 1943.

[310] L. R. G. Treloar. Stress-strain data for vulcanized rubber under various types of deformation. *Transactions of the Faraday Society*, 40:59–70, 1944.

[311] L. R. G. Treloar. *The Physics of Rubber Elasticity*. Oxford University Press, 3rd edition, 2005.

[312] C. Truesdell and W. Noll. *The Non-Linear Field Theories of Mechanics*. Springer-Verlag, Berlin, 3rd edition, 2004.

[313] R. Vaishnav and J. Vossoughi. *Biomedical Engineering II, Recent Developments*, chapter Estimation of residual strain in aortic segments. Pergamon Press, 1983.

[314] E. van Dam, S. Dams, G. Peters, M. Rutten, G. Schurink, J. Buth, and F. van de Vosse. Non-linear viscoelastic behavior of abdominal aortic aneurysm thrombus. *Biomechanics and Modeling in Mechanobiology*, 7(2):127–137, 2008.

[315] F. van de Vosse, J. de Hart, C. van Oijen, D. Bessems, T. Gunther, A. Segal, B. Wolters, J. Stijnen, and F. Baaijens. Finite-element-based computational methods for cardiovascular fluid-structure interaction. *Journal of Engineering Mathematics*, 47(3):335–368, 2003.

[316] J. P. Vande Geest, M. S. Sacks, and D. A. Vorp. A planar biaxial constitutive relation for the luminal layer of intra-luminal thrombus in abdominal aortic aneurysms. *Journal of Biomechanics*, 39(13):2347–2354, 2006.

[317] M. Vianello. Coaxiality of strain and stress in anisotropic linear elasticity. *Journal of Elasticity*, 42(3):283–289, 1996.

[318] I. E. Vignon-Clementel, A. C. Figueroa, K. E. Jansen, and C. A. Taylor. Outflow boundary conditions for three-dimensional finite element modeling of blood flow and pressure in arteries. *Computer Methods in Applied Mechanics and Engineering*, 195(29-32):3776–3796, 2006.

[319] R. P. Vito and S. A. Dixon. Blood vessel constitutive models — 1995–2002. *Annual Review of Biomedical Engineering*, 5(1):413–439, 2003.

[320] D. A. Vorp. Biomechanics of abdominal aortic aneurysm. *Journal of Biomechanics*, 40(9):1887–1902, 2007.

[321] D. A. Vorp, W. A. Mandarino, M. W. Webster, and J. Gorcsan. Potential influence of intraluminal thrombus on abdominal aortic aneurysm as assessed by a new non-invasive method. *Cardiovascular Surgery*, 4(6):732–739, 1996.

[322] L. Vu-Quoc and X. G. Tan. Optimal solid shells for non-linear analyses of multilayer composites. I. Statics. *Computer Methods in Applied Mechanics and Engineering*, 192(9-10):975–1016, 2003.

[323] L. Vu-Quoc and X. G. Tan. Optimal solid shells for non-linear analyses of multilayer composites. II. Dynamics. *Computer Methods in Applied Mechanics and Engineering*, 192(9-10):1017–1059, 2003.

[324] W. A. Wall. *Fluid-Struktur-Interaktion mit stabilisierten Finiten Elementen*. PhD thesis, Institut für Baustatik, Universität Stuttgart, 1999.

[325] W. A. Wall, M. Bischoff, and E. Ramm. A deformation dependent stabilization technique, exemplified by EAS elements at large strains. *Computer Methods in Applied Mechanics and Engineering*, 188(4):859–871, 2000.

[326] W. A. Wall, M. A. Frenzel, and C. Cyron. Isogeometric structural shape optimization. *Computer Methods in Applied Mechanics and Engineering*, 197(33-40):2976–2988, 2008.

[327] D. H. J. Wang, M. Makaroun, M. W. Webster, and D. A. Vorp. Mechanical properties and microstructure of intraluminal thrombus from abdominal aortic aneurysm. *Journal of Biomechanical Engineering*, 123(6):536–539, 2001.

[328] M. C. Wang and E. Guth. Statistical theory of networks of non-gaussian flexible chains. *Journal of Chemical Physics*, 20(7):1144–1157, 1952.

[329] P. Watton and N. Hill. Evolving mechanical properties of a model of abdominal aortic aneurysm. *Biomechanics and Modeling in Mechanobiology*, 8(1):25–42, 2009.

[330] P. Watton, N. Hill, and M. Heil. A mathematical model for the growth of the abdominal aortic aneurysm. *Biomechanics and Modeling in Mechanobiology*, 3(2):98–113, 2004.

[331] E. L. Wilson, R. L. Taylor, W. P. Doherty, and J. Ghabouss. *Numerical and Computer Models in Structural Mechanics*, chapter Incompatible displacement models, pages 43–57. Academic Press, New York, 1973.

[332] B. Wolters, M. Rutten, G. Schurink, U. Kose, J. de Hart, and F. van de Vosse. A patient-specific computational model of fluid-structure interaction in abdominal aortic aneurysms. *Medical Engineering & Physics*, 27(10):871–883, 2005.

[333] P. Wriggers. *Nichtlineare Finite-Element-Methoden*. Springer, Berlin, 2001.

[334] P. Wriggers and F. Gruttmann. Thin shells with finite rotations formulated in biot stresses: Theory and finite element formulation. *International Journal for Numerical Methods in Engineering*, 36(12):2049–2071, 1993.

[335] P. Wriggers and S. Reese. A note on enhanced strain methods for large deformations. *Computer Methods in Applied Mechanics and Engineering*, 135(3-4):201–209, 1996.

[336] W. Wu, W.-Q. Wang, D.-Z. Yang, and M. Qi. Stent expansion in curved vessel and their interactions: A finite element analysis. *Journal of Biomechanics*, 40(11):2580–2585, 2007.

[337] F. L. Wuyts, V. J. Vanhuyse, G. J. Langewouters, W. F. Decraemer, E. R. Raman, and S. Buyle. Elastic properties of human aortas in relation to age and atherosclerosis: a structural model. *Physics in Medicine and Biology*, 10(10):1577, 1995.

[338] C. K. Zarins, D. P. Giddens, B. K. Bharadvaj, V. S. Sottiurai, R. F. Mabon, and S. Glagov. Carotid bifurcation atherosclerosis quantitative correlation of plaque localization with flow velocity profiles and wall shear-stress. *Circulation Research*, 53(4):502–514, 1983.

[339] Y. Zhang, Y. Bazilevs, S. Goswami, C. L. Bajaj, and T. J. R. Hughes. Patient-specific vascular NURBS modeling for isogeometric analysis of blood flow. *Computer Methods in Applied Mechanics and Engineering*, 196(29-30):2943–2959, 2007.

[340] Y. Zhang, M. Dunn, E. Drexler, C. McCowan, A. Slifka, D. Ivy, and R. Shandas. A microstructural hyperelastic model of pulmonary arteries under normo- and hypertensive conditions. *Annals of Biomedical Engineering*, 33(8):1042–1052, 2005.

[341] O. C. Zienkiewicz and R. L. Taylor. *The Finite Element Method, Vol. 2 - Solid Mechanics*. Butterworth-Heinemann, 5 edition, 2000.

[342] O. C. Zienkiewicz, R. L. Taylor, and J. Z. Zhu. *The Finite Element Method - Its Basis and Fundamentals*. Butterworth-Heinemann, 6th edition, 2005.

[343] M. A. Zulliger, P. Fridez, K. Hayashi, and N. Stergiopulos. A strain energy function for arteries accounting for wall composition and structure. *Journal of Biomechanics*, 37(7):989–1000, 2004.

Die VDM Verlagsservicegesellschaft sucht für wissenschaftliche Verlage abgeschlossene und herausragende

Dissertationen, Habilitationen, Diplomarbeiten, Master Theses, Magisterarbeiten usw.

für die kostenlose Publikation als Fachbuch.

Sie verfügen über eine Arbeit, die hohen inhaltlichen und formalen Ansprüchen genügt, und haben Interesse an einer honorarvergüteten Publikation?

Dann senden Sie bitte erste Informationen über sich und Ihre Arbeit per Email an *info@vdm-vsg.de*.

Sie erhalten kurzfristig unser Feedback!

VDM Verlagsservicegesellschaft mbH
Dudweiler Landstr. 99 Telefon +49 681 3720 174
D - 66123 Saarbrücken Fax   +49 681 3720 1749
www.vdm-vsg.de

Die VDM Verlagsservicegesellschaft mbH vertritt

Printed by Books on Demand GmbH, Norderstedt / Germany